Alan Ellis
Liz Highleyman
Kevin Schaub
Melissa White
Editors

The Harvey Milk Institute Guide to Lesbian, Gay, Bisexual, Transgender, and Queer Internet Research

Pre-publication
REVIEWS,
COMMENTARIES,
EVALUATIONS . . .

"*The HMI Guide to LGBTQ Internet Research,* designed to become your most tattered and well-used book, should be placed next to your computer for easy and frequent access. The *HMI Guide* is exactly what those of us who conduct research on LGBTQ issues or who want to educate ourselves about such matters have been waiting for—expert advice on how best to navigate the Web. There is history, suggestions on how to use the Web for research and education, sage advice from those who have made LGBTQ information available to us on the Internet, explanations for the beginner and the veteran, and every Internet address you will need—made accessible within the confines of this single book, with assurances of periodic online updates. Online resources for the social and biological sciences, medicine, the humanities, the arts, business, law, media, queer studies, and community service—it is all here. An extraordinarily valuable guide that will save you hours of Web surfing. It will be one of the most functional and essential acquisitions you make this year."

Ritch C. Savin-Williams, PhD
Professor of Developmental and Clinical Psychology,
Cornell University,
Ithaca, NY

More pre-publication
REVIEWS, COMMENTARIES, EVALUATIONS . . .

"The *Guide to LGBTQ Internet Research* is a timely reference tool that will be useful to students, journalists, and researchers in a surprisingly comprehensive range of fields.

In the fast-changing world of online research, the *Guide* emphasizes well-established and institutional resources. The interviews with pathbreakers, founders, and leaders of Internet research organizations will give the *Guide* a long shelf life."

Kevin G. Barnhurst, PhD
Associate Professor,
University of Illinois at Chicago

"At last, a book to help researchers navigate the LGBTQ World Wide Web! With so many Web sites and so much information available on the Web today, it can be difficult for even seasoned researchers to know where to go to find reliable, timely, and helpful information. This book simplifies that task with its explanations of the strengths of different search engines and portals, and suggests Web sites for information in a variety of fields—from anthropology to economics to transgender and intersex research.

This book is an indispensable aid to students interested in LGBTQ topics and courses. It is useful for both researchers and 'students of life' looking for information. There are a lot of us who use the Web every day at work and at home who will benefit from the efficiency of having a resource available to help steer us in the right direction as we search the Web for information. I am already using the book for my research, my teaching, and for general information surfing."

Ellen D. B. Riggle, PhD
Associate Professor,
Department of Political Science;
Associate Director,
Women's Studies Program,
University of Kentucky, Lexington

Harrington Park Press®
An Imprint of The Haworth Press
New York • London • Oxford

NOTES FOR PROFESSIONAL LIBRARIANS AND LIBRARY USERS

This is an original book title published by Harrington Park Press, an imprint of The Haworth Press, Inc. Unless otherwise noted in specific chapters with attribution, materials in this book have not been previously published elsewhere in any format or language.

CONSERVATION AND PRESERVATION NOTES

All books published by The Haworth Press, Inc. and its imprints are printed on certified pH neutral, acid free book grade paper. This paper meets the minimum requirements of American National Standard for Information Sciences-Permanence of Paper for Printed Material, ANSI Z39.48-1984.

The Harvey Milk Institute Guide to Lesbian, Gay, Bisexual, Transgender, and Queer Internet Research

HAWORTH Gay & Lesbian Studies
John P. De Cecco, PhD
Editor in Chief

Behold the Man: The Hype and Selling of Male Beauty in Media and Culture by Edisol Wayne Dotson

Untold Millions: Secret Truths About Marketing to Gay and Lesbian Consumers by Grant Lukenbill

It's a Queer World: Deviant Adventures in Pop Culture by Mark Simpson

In Your Face: Stories from the Lives of Queer Youth by Mary L. Gray

Military Trade by Steven Zeeland

Longtime Companions: Autobiographies of Gay Male Fidelity by Alfred Lees and Ronald Nelson

From Toads to Queens: Transvestism in a Latin American Setting by Jacobo Schifter

The Construction of Attitudes Toward Lesbians and Gay Men edited by Lynn Pardie and Tracy Luchetta

Lesbian Epiphanies: Women Coming Out in Later Life by Karol L. Jensen

Smearing the Queer: Medical Bias in the Health Care of Gay Men by Michael Scarce

Macho Love: Sex Behind Bars in Central America by Jacobo Schifter

When It's Time to Leave Your Lover: A Guide for Gay Men by Neil Kaminsky

Strategic Sex: Why They Won't Keep It in the Bedroom edited by D. Travers Scott

One of the Boys: Masculinity, Homophobia, and Modern Manhood by David Plummer

Homosexual Rites of Passage: A Road to Visibility and Validation by Marie Mohler

Male Lust: Pleasure, Power, and Transformation edited by Kerwin Kay, Jill Nagle, and Baruch Gould

Tricks and Treats: Sex Workers Write About Their Clients edited by Matt Bernstein Sycamore

A Sea of Stories: The Shaping Power of Narrative in Gay and Lesbian Cultures—A Festschrift for John P. De Cecco edited by Sonya Jones

Out of the Twilight: Fathers of Gay Men Speak by Andrew R. Gottlieb

The Mentor: A Memoir of Friendship and Gay Identity by Jay Quinn

Male to Male: Sexual Feeling Across the Boundaries of Identity by Edward J. Tejirian

Straight Talk About Gays in the Workplace, Second Edition by Liz Winfeld and Susan Spielman

The Bear Book II: Further Readings in the History and Evolution of a Gay Male Subculture edited by Les Wright

Gay Men at Midlife: Age Before Beauty by Alan L. Ellis

Being Gay and Lesbian in a Catholic High School: Beyond the Uniform by Michael Maher

Finding a Lover for Life: A Gay Man's Guide to Finding a Lasting Relationship by David Price

The Man Who Was a Woman and Other Queer Tales from Hindu Lore by Devdutt Pattanaik

How Homophobia Hurts Children: Nurturing Diversity at Home, at School, and in the Community by Jean M. Baker

The Harvey Milk Institute Guide to Lesbian, Gay, Bisexual, Transgender, and Queer Internet Research edited by Alan Ellis, Liz Highleyman, Kevin Schaub, and Melissa White

Stories of Gay and Lesbian Immigration: Together Forever? by John Hart

The Harvey Milk Institute Guide to Lesbian, Gay, Bisexual, Transgender, and Queer Internet Research

Alan Ellis
Liz Highleyman
Kevin Schaub
Melissa White
Editors

HPP

Harrington Park Press®
An Imprint of The Haworth Press
New York • London • Oxford

Published by

Harrington Park Press®, an imprint of The Haworth Press, Inc., 10 Alice Street, Binghamton, NY 13904-1580.

Cover design by Evan Deerfield.

Library of Congress Cataloging-in-Publication Data

The Harvey Milk Institute guide to lesbian, gay, bisexual, transgender, and queer internet research / Alan Ellis ... [et al.], editors.
p. cm.
Includes bibliographical references and index.
ISBN 1-56023-352-4 (alk. paper)—ISBN 1-56023-353-2 (alk. paper)
1. Homosexuality—Computer network resources. 2. Gays—Computer network resources. I. Ellis, Alan, 1957- II. Harvey Milk Institute.

HQ76.25 .H375 2001
025.04'086'64—dc21

2001039707

CONTENTS

About the Editors ix

Contributors xi

Foreword xiii
Ronni Sanlo

Preface xvii

Chapter 1. Introduction 1
Alan Ellis
Kevin Schaub
Melissa White

Using This Guide 2
A Note on Comprehensiveness 3

Chapter 2. Conducting Research on the Internet 5
Liz Highleyman
Warren Longmire
Sanda Steinbauer

Search Engines and Directories 6
Non-Web Resources 11
Credibility 16
Citations 18

Chapter 3. Major LGBTQ Internet Research Tools 23
Alan Ellis
Liz Highleyman

Major LGBTQ Portals 23
Online Directories and Other Web Sites 25
Online Database of Journal Articles 28
Major LGBTQ Non-Web Resources 28

Interview: Mark Elderkin and Rhona Berenstein 30
Interview: Megan Smith 34
Interview: Roger Klorese and Will Doherty 44

Chapter 4. Queer Studies **49**
Sanda Steinbauer

Queer Studies Resources 50
Non-Web Resources 54

Chapter 5. Bisexual Studies **55**
Liz Highleyman

Bisexual Studies Resources 55
Non-Web Resources 58

Chapter 6. Transgender and Intersex Studies **61**
Alan Ellis
Liz Highleyman

Transgender and Intersex Studies Resources 61
Non-Web Resources 65
Interview: Gwendolyn Ann Smith 66

Chapter 7. Human Sexuality Studies **71**
Liz Highleyman
Sanda Steinbauer

General Sites 72
BDSM/Leather/Fetish Resources 75
Polyamory Resources 77
Sexual Health and Education Sites 78
Sex Work Resources 79
Non-Web Resources 80

Chapter 8. Liberal Arts and the Humanities **83**
Alan Ellis
Mark Menke

History 83
Ethnic Studies 87
Religious Studies 89
Non-Web Resources 98

Chapter 9. Social and Biological Sciences **101**
Alan Ellis

General Sites 101
Anthropology 102
Political Science 103
Psychology 105
Sociology 108
Biological Sciences 109
Non-Web Resources 110
Interview: Ellen D. B. Riggle 110

Chapter 10. Arts and Education **115**
Alan Ellis

Arts 115
Education 117
Interview: Barry Harrison 117

Chapter 11. Law and Philosophy **127**
Alan Ellis

Law 127
Philosophy 129

Chapter 12. Health and Medicine **131**
Liz Highleyman

General LGBTQ Health Issues 131
Gay Men's Health Issues 133
Lesbian Health Issues 134
Bisexual Health Issues 136
Transgender and Intersex Health Issues 137
HIV/AIDS Sites 138
General Consumer Health Sites 140
Researching the Medical Literature 142
Non-Web Resources 142
Interview: Sister Mary Elizabeth 143

Chapter 13. Business, Labor Studies, and Economics **147**
Alan Ellis

General Sites 147
Labor Studies and Workplace Issues 148
Domestic Partnership and Workplace Nondiscrimination Issues 149
Economics 150
Non-Web Resources 151

Chapter 14. Community Resources **153**
Melissa White

LGBTQ Community-Based Resources 153
Non-Web Resources 156
Interview: Kevin Schaub 156

Chapter 15. Media and News **161**
Mark Menke

Media Resources 161
Non-Web Resources 163

Chapter 16. The Queer Internet: One Student's Experience **165**
David Brightman

Index **169**

ABOUT THE EDITORS

Alan L. Ellis, PhD, is co-author of several books on LGBTQ topics, including *Gay Men at Midlife: Age Before Beauty* (Haworth, 2001), *Sexual Identity on the Job: Issues and Services* (Haworth, 1996), and *A Family and Friend's Guide to Sexual Orientation*. He is co-chair of the Board of Directors of the Harvey Milk Institute. He earned a PhD in psychology from the University of Illinois and teaches at San Francisco State University.

Liz Highleyman, CPH, is a freelance writer, editor, and health educator. She worked at the Massachusetts Institute of Technology's Artificial Intelligence Laboratory from the mid-1980s to the mid-1990s, where she witnessed the birth of the World Wide Web. She has worked as Health Editor for the Internet search engine Ask Jeeves and Acting Editor of the Bulletin of Experimental Treatment for AIDS. She co-edited the anthology *Bisexual Politics: Theories, Queries, and Visions* (Haworth), and her work has appeared in the *Encyclopedia of AIDS, The Second Coming,* and *Bi Any Other Name*. Liz has a BA in cognitive science from the University of Rochester and a certification in public health from the Harvard School of Public Health.

Kevin Schaub, MA, is the Executive Director and Dean of the Harvey Milk Institute, the world's largest center for queer studies. He serves as the Festival Director for the National Queer Arts Festival each summer. He holds an MA in modern French literature and cinema from The Ohio State University and a BA in French and political science from Rutgers University. Additionally, Kevin has worked as an editor, as a translator/interpreter, and as an instructor in language, popular culture, and cinema studies.

Melissa White, MFA, is a queer bisexual activist, writing teacher, and nonprofit fundraiser. She works with the Harvey Milk Institute in each of these capacities and also serves as co-chair of the Board of Directors. Melissa, who earned her MFA in writing from Mills College, is pleased to be co-editor of a forthcoming book about social and political aspects of bisexuality, a project of the Policy Institute of the National Gay and Lesbian Task Force.

CONTRIBUTORS

David Brightman is a San Francisco-based writer who works in acquisitions at Jossey-Bass Publishers. He has been a big homo his whole life, and his queer interests received academic validation through interdisciplinary study of gender and sexuality at the University of California at Berkeley (1994-1996). He currently lives a sedate but still queer life in San Francisco with his partner and their small dog.

Warren Longmire designs and develops online education programs at Viviance New Education in San Francisco. He is co-author of *Managing Web-Based Training* (ASTD, 1999) and is a contributor to *Web-Based Instruction: Lessons from the Pro*s (Educational Technology Publications, 2001). He serves on the advisory board of the Harvey Milk Institute and volunteers with the San Francisco Bicycle Coalition.

Mark Menke is program director of the Harvey Milk Institute. He holds a BA in journalism from Marquette University, a Jesuit college in Milwaukee, WI. As a former co-facilitator of the Catholic, conservative school's GLB student group, he has an interest in the intersection of one's faith/religion/spirituality and one's sexual and gender identity.

Mouwafa Sidaoui is a manager of technical training and curriculum development in Silicon Valley. He co-edited *Statistical Explorations with Microsoft Excel* (Duxbury), and has taught computer and management science at the University of San Francisco and Golden Gate University. Mouwafa is originally from Lebanon and moved to the United States at age eighteen. He received his BS in mechanical engineering and engineering management and holds an MBA from Boston University.

Sanda Steinbauer is a freelance writer, scholar, and activist. She earned her MA from New York University in Performance Studies and currently resides in San Francisco and New York City.

Elizabeth Taylor works as a Silicon Valley senior documentation architecture and engineering consultant for corporations such as Hewlett-Packard, National Semiconductor, ATI, and a number of Internet startups. She is member of the Queer Arts Resource board of directors and has taught Advanced Technical Communications in the Engineering Department of the University of California at Berkeley Extension Program. She has worked as the Managing Editor of the National Women's Studies Association (NWSA) academic journal and has taught literature and writing courses at the university level. She has an MA in English literature from Indiana University in Bloomington, Indiana, and a PhD (pending, ABD) from Ohio State University in Columbus, Ohio. For fun, she writes fiction, oil paints, studies classical guitar and digital recording, and trains for a soccer team competing in the 2002 Gay Games in Australia.

Foreword

I wanted to do my dissertation on sexual orientation issues in the residence halls where I work at my university but my committee chair said the topic is too narrow. I have to find something else. But I'm a gay man and this is work of my heart. There's so much I want to know. I have this opportunity to do this work and my chair said no.

Imagine a doctoral student wishing to write a dissertation on a particular aspect of people with disabilities. Imagine he or she is told by the committee chair that the topic is too narrow, that it will not be taken seriously, and that there is not enough grounded research in that area to support the student's work. Imagine a faculty member who wants to research some aspect of an ethnic identity and is told by the department chair that the area is too limited and therefore it is not fundable, and that because the topic is not considered important by the department there will be no release time to do such research. These are preposterous responses to researchers wishing to explore aspects of the world that are important to them. Topic too narrow? Not taken seriously? No funding possibilities? No release time? And yet people who wish to embark on research about lesbian, gay, bisexual, and transgender (LGBT) people hear these statements regularly, even at research I institutions in the United States: the topic of LGBT is too narrow, there is no grounded research, there is no funding to support one's efforts.

As an educator, a researcher, and an administrator at a research I institution, I wish I could tell you that it is different at my university. It's not. The difference at UCLA is that people like Allison Diamant, Vernon Rosario, Anne Peplau, Susan Cochran, Vickie Mays, Linda Garnets, myself, and others are driven to do this work regardless of the level of support we receive from our departments. Much of our work is done on our own time and with our own money. However, some of us are beginning to seek out, locate, and actually receive funding to explore LGBT issues with at least a base of financial support. The Lesbian Health Fund of the Gay and Lesbian Medical Association and a few of our national professional organizations are beginning to assist with much needed funding so that researchers are able to do the work, to broaden the topic area of LGBT people and issues, and to have LGBT areas of research be taken seriously by our departments. For the first time, in May 2001, a major federal agency, the National Institutes of Health, released a program announcement calling for research proposals that examine the areas of basic, clinical intervention, practice, and services research in LGBT issue areas. They ask that projects "make use of the most rigorous current methodologies and, where needed, engage in development and evaluation of new methodologies." Finally our work and our issues will receive the attention they deserve as valued research topic areas to be taken seriously by higher education institutions. Now the challenge becomes not IF we can do the research but HOW? How do we locate the few pieces—in the big scheme of the vast arena of research—that may be available?

One of the reasons cited by departments when they discourage LGBT research is that there is not enough grounded work in this field. There are, indeed, many articles and books about LGBT people and issues but there is still not yet a plethora of empirical data or vigorous qualitative work. There has not been an easy or efficient way to access the work that is available beyond usual library searches.

The World Wide Web provides enormous opportunities for researching LGBT topic areas. College students in the twenty-first century are more than Web-savvy; they are Web-educated. The Web has become their primary, or at least, initial context of information, regardless of the topic. The diversity of LGBT subjects is vast and many sites on the Web may or may not be helpful as students and faculty embark on their research, whether the areas are in queer theory/studies, or in the social or hard sciences. But once one knows where and how to look on the Web, the universe expands dramatically. The topic of LGBT people and issues suddenly becomes less narrow, less hard to find, and more available for expansion to the grounded work so dearly needed.

As significant academic research into the historical, social, policy, political, and organizational issues that impact and are impacted by LGBT people increases, all areas of research expand exponentially. Gender, race, sexual identity, family, and so much more may be subcategorized within each of the previously listed issues. This book provides the foundational information and access to materials and information in every discipline. Never again should we hear that queer topics are too narrow. Alan Ellis and the contributors to this work have opened the door to the universe. This is one researcher who is thrilled!

Ronni Sanlo, PhD
Director of UCLA LGBT
Campus Resource Center
Los Angeles, California

Preface

This guide was prepared by volunteers and staff of the Harvey Milk Institute (HMI) in San Francisco. HMI is the world's largest community-based center for lesbian, gay, bisexual, transgender, and queer studies. The institute partners with organizations throughout the San Francisco Bay Area and beyond to offer opportunities to examine and create queer culture.

Founded in 1994, HMI is a nonprofit 501(c)3, community-based institution that works to fulfill the goals of San Francisco Supervisor Harvey Milk. Harvey Milk won a seat on the San Francisco Board of Supervisors in 1977, becoming one of the first openly gay elected officials in the United States. Through tireless coalition building, campaigning, and good humor, Milk successfully integrated people and politics across diverse boundaries. Tragically, after serving only eleven months, and at the peak of his influence in city, state, and national civil rights, Milk was assassinated along with Mayor George Moscone by Supervisor Dan White on November 27, 1978. (For an excellent documentary on the life of Harvey Milk, see *Times of Harvey Milk,* narrated by Harvey Fierstein and produced by Rob Epstein.)

To find out more about HMI, go to <http://www.harveymilk.org>.

Chapter 1

Introduction

Alan Ellis
Kevin Schaub
Melissa White

Not long ago, choosing a research topic that focused on lesbian, gay, bisexual, transgender, or queer (LGBTQ)* issues was an act of courage and activism. Even today, many universities, colleges, and libraries refuse to support queer researchers by failing to offer resources such as books and journals that address gender and queer studies. Although the increasing acceptance of queer studies as a legitimate academic pursuit has helped improve the overall climate and available resources for queer researchers—especially at larger universities—queer studies is often marginalized. The marginalization of the discipline means that many institutions still offer only limited resources. Fortunately, the development of queer resources on the Internet means that students and researchers have access to queer research materials, regardless of what may or may not be physically present on their campuses or in their local libraries.

The purpose of this guide is to help researchers understand how to access and use the many resources on the Internet that are specifically useful for academic research on queer topics in a variety of disci-

*As of yet, there is no universally accepted acronym for the queer community, and the diversity of acronyms used in writings by and about the queer community reflects underlying differences in our politics and regional traditions. Throughout this guide, we have chosen to use the acronym LGBT with the addition of a "Q" to refer to those who identify as queer or questioning. However, in some cases other acronyms will appear (e.g., GLBT or LBT) based on usage within and relevance to a given Web site or Internet resource.

plines, including the social sciences, business, the humanities, the arts, and, of course, queer studies.

The guide also includes interviews with people who have played a key role in offering information and other resources on LGBTQ topics on the Internet. These individuals include the co-founder and president of Gay.com, the CEO of PlanetOut, the founder of QueerNet, the Webmistress of Gender.org, the executive director of one of the key Web sites for queer arts (Queer Arts Resource), a professor of political science, and the founder of AEGiS (an HIV/AIDS Internet resource).

USING THIS GUIDE

The Harvey Milk Institute Guide to Lesbian, Gay, Bisexual, Transgender, and Queer Internet Research includes uniform resource locators (URLs) that provide the Internet addresses for online resources. These URLs are arranged by general academic hierarchies, such as discipline and subject area. For example, sites of particular interest to someone conducting anthropological research are found in the social sciences chapter in the section on anthropology. In each section, the Web sites are arranged alphabetically. Wherever possible, the guide also includes non-Web resources such as Usenet newsgroups and electronic mailing lists. You can find a description of these resources and how to use them in Chapter 2, "Conducting Research on the Internet."

Regardless of your research area of interest, the two major LGBTQ portals—Gay.com and PlanetOut—the Queer Resources Directory, QueerNet, and Homorama are excellent resources (described in Chapter 3, "Major LGBTQ Internet Research Tools)." These and other online resources are mentioned more than once in this guide, when specific pages or links are relevant to a certain discipline or subject area. However, these resources may be useful for subject areas where they are not listed as well.

To help you access the various Web sites and URLs listed in this guide without having to type in each URL, go to <http://www.harveymilk.org> and click on the "Queer Internet Research Guide" link. You will find all of the URLs listed in this guide, arranged by chapter. We will update the links as needed.

A NOTE ON COMPREHENSIVENESS

This guide does not—of course—list every useful resource on the Internet for LGBTQ research. In most cases, in an effort to minimize the number of dead links, the contributors have selected large, well-established, English-language sites that are likely to be maintained well into the future. Inevitably, we will have missed some important resources. Also, you will notice a preponderance of resources maintained by U.S. institutions, scholars, and activists among these listings. This bias toward U.S.-based resources reflects our knowledge and experience with the LGBTQ Internet, and the limitations of that knowledge. We encourage readers to recommend to us additional useful sites that are not included in this guide, as well as notifying us of any links that no longer function. (Send suggested URLs and other suggestions to <harvmilk@aol.com>.)

Also, for certain disciplines, we have listed print resources. Generally, we did this only when there were few online resources for the discipline. Where online resources were plentiful, we expect that you will find many references to print resources online.

Chapter 2

Conducting Research on the Internet

Liz Highleyman
Warren Longmire
Sanda Steinbauer

This chapter focuses on how to use Web search engines and directories, how to access non-Web resources such as mailing lists and Usenet newsgroups, how to assess the credibility of the information that you find on the Internet, and how to reference online materials in your work.

The Internet is a global network of interconnected computers that use a set of shared protocols to "talk" to one another. The term "information superhighway" is fortuitous, since the Internet can be likened to a road system with large freeways connecting major hubs, branching into smaller streets that lead to individual homes.

Originally known as ARPAnet, the Internet began in the late 1960s as a project of the U.S. Department of Defense, in an effort to create a decentralized system that could survive a disaster. In the mid-1980s, the National Science Foundation created NSFnet, an expanded network for research and educational institutions. Throughout the 1980s, the Internet was primarily used by researchers, students, and others with access to university computers. As such, it was mainly used for free information exchange, be it scholarly articles or aimless banter.

For information about the development of the LGBTQ community on the Internet, see the interview with Megan Smith in Chapter 3, "Major LGBTQ Internet Research Tools." For a general overview of the Internet, its history, and how to use it, see the *EFF's (Extended) Guide to the Internet* (http://www.eff.org/pub/Net_info/ EFF_Net_Guide/netguide.eff) published by the Electronic Frontier Foundation.

SEARCH ENGINES AND DIRECTORIES

The Web was developed by Tim Berners-Lee and colleagues in the early 1990s, and graphical Web browsers became widely available in the mid-1990s, kicking off an explosion of growth in Internet use. At the end of 2000, it was estimated that nearly 400 million people worldwide used the Web, and many of these users find search engines and directories to be among their most valuable tools. As the Web has grown from a few sites to millions or billions, though, combing through the results returned by a search engine can feel like taking a drink from a fire hose. For example, entering the word "gay" into Altavista returns over 3,000,000 results.

A search engine, in the most basic terms, is a site that returns lists of matching URLs (usually with a brief description) when you enter keywords or phrases. Search engines typically let you use Boolean operators (AND, OR), adjacency delimiters, quotes, domain name specifications, wildcards, or other criteria to refine and narrow your searches. Fortunately, all the major search engines provide a "how to" section that explains how to get the best results from that site's search technology.

Most search engines use programs called "spiders" that go out and survey available Web sites. Some look at all the text on a Web page, while others focus on the text contained in "meta tags" in a page's HTML code. Meta tags are supposed to describe what a page is about; however, Web page creators can put whatever they want in their metatags, and some place common search words—such as "sex" and "Viagra"—in their tags to drive more traffic, even though a site may have nothing whatsoever to do with sex or Viagra. Unfortunately, the terms "gay" and especially "lesbian" have been used in this manner, and it's not uncommon for a search using these terms to return a plethora of pornography sites.

To get around such problems, search engine developers are continually developing new schemes to try to ensure that search results are as relevant as possible. Some search engines use algorithms that rank their search results by popularity and usefulness. For example, Direct Hit rates sites based on how many other users visit a site. Google ranks its results based on how often Webmasters include links to a given site on their own pages, under the assumption that the pages that are linked to most often are the most useful.

Directories are listings of Web links categorized into a topic-based directory structure. Instead of entering keywords, you navigate through a directory tree, moving from more specific to more narrow topics until you find what you're looking for. It can often be difficult to find queer resources in the large directories, since some tend to bury so-called controversial material in an effort to be "family friendly." In general, queer content can be found under top-level topic headings such as "Society," "Culture," "Lifestyle," or "People."

Today, most of the largest and best-known search sites offer both a search engine and a directory, often along with other features such as news headlines, yellow pages, chat areas, and shopping. These sites, commonly known as "portals," can be good starting points for research, especially in an area that is new to you. Even if the results provided are too general, these sites can often guide you in the right direction. Because search engines work in different ways and can give varying results, some users prefer "metasearches," e.g., sites such as Dogpile that return results from multiple search engines with a single search.

The Internet explosion, not surprisingly, has been seen as a huge market opportunity. In an ever more competitive commercial environment, some search engines and directories have taken steps to "monetize" their services, for example by letting companies pay for more rapid review of their sites and even for better placement in the list of search results. This phenomenon, along with the growth in the sheer number of e-commerce sites, has made Web sites that simply provide free information more difficult to find. They're still out there, of course, but can easily get lost in the flood of sites trying to sell you something.

There are too many search engines and directories to list comprehensively; think of the links that follow as a representative sample.

Altavista
http://www.altavista.com

Altavista is one of the earliest and largest search engines. It returns a huge number of seemingly randomly organized results for general searches (e.g., "gay"), many of which are not highly relevant. Unlike the other search engines listed in this section, you must use Boolean operators or quotes to get reasonable results for multiword searches (e.g., queer AND studies, or "queer studies," but not queer studies).

Altavista provides a directory based on the Open Directory Project classification scheme (see Open Directory Project). One popular feature of Altavista is its ability to translate Web sites into different languages.

Ask Jeeves
http://www.ask.com

Ask Jeeves is a question-answering and search site that lets users enter questions in a natural language format and returns answers selected by human editors. Content is good in some areas and lacking in others. The site also returns popularity-based results provided by Direct Hit (see Direct Hit), and provides access to a directory based on the Open Directory Project classification scheme.

Direct Hit
http://www.directhit.com

Direct Hit returns search results ranked by popularity. Sometimes results can be odd; for example, when searching on the word "gay," personal and local sites are ranked almost as high as Gay.com, one of the two largest gay portals. Performance is better for more specific searches. Direct Hit also features a directory based on the Open Directory Project scheme and includes suggestions for related searches.

Dogpile
http://www.dogpile.com

Dogpile is a metasearch that returns results from several of the most popular search engines. Metasearches can be a good way to get a broad overview of a topic, because you are not limited by the peculiarities of a single search engine. Dogpile provides access to LookSmart's directory (see LookSmart). To find queer resources, select Library, then Society, then Gay and Lesbian.

Excite
http://www.excite.com

Excite is a full-service portal with a rather cluttered interface. It returns relevant results for both general and specific searches. For example, entering "gay" returns URLs for the Web sites of several major gay organizations and portals. The "Explore Excite" section on the home page looks like a directory of links, but is actually a collection of message boards, chat forums, and other various features.

Go Network
http://www.go.com

The Go Network (including the former Infoseek) features both a search engine and a directory. The links on the home page lead to "channels"; to find the Go Guides directory, click the Search tab at the top of the page. To find queer resources, select People, then Gay Community. From here, there are many subsections, including those for bisexuals, transgendered people, lesbians, queer teens, and seniors. There is also a listing of gay search engines. Go.com tends to return different results than other search engines, so it can be a good place to find information that's off the beaten path.

Google
http://www.google.com

Google is a fast-growing, second generation search engine that many users appreciate for its uncluttered interface. Google's technology sorts URLs by relevance based on the prominence of keywords on a page and by how often a given page is linked to from other sites. Both general and specific searches produce useful results. Google has become one of the largest and most popular search engines, and indexes over a billion Web pages. The site also includes a directory based on the Open Directory Project classification scheme.

LookSmart
http://www.looksmart.com

The LookSmart directory includes many of the best and most well-known links in a given topic area. To find queer resources, select Library, then Society, then Gay and Lesbian. From here, there are subsections such as education, health and fitness, history, and society and politics. Links for lesbians, bisexuals, transgendered people, queers of various ethnicities, youth, and families are included in the Community and Culture section; this section also includes subsections for queer organizations, magazines, and literature.

Lycos
http://www.lycos.com

Lycos is another large portal featuring both a search engine and a directory based on the Open Directory Project classification scheme. General searches do not return results sorted by relevance, although more specific searches tend to return more useful results.

Northern Light
http://www.northernlight.com

Northern Light is a relatively new search engine with several interesting features. It provides simple and "power" search capabilities, and also lets you search recent news headlines. You can limit your search to specific subject areas (for example, arts, humanities, or reference) or document type (for example, government Web sites, nonprofit Web sites, personal Web pages, or college newspapers). You can request that your search results be sorted by relevance or date.

Open Directory Project
http://www.dmoz.org

The Open Directory Project (also known as DMOZ, after Netscape's lizard mascot, Mozilla), was developed using the open source model. Web site listings are compiled by a large group of volunteer editors with expertise or interest in a given topic area. The project's goal is to allow users to organize the Internet as they see fit, and the Open Directory Project classification scheme is now used by many search engines. As a volunteer initiative, the project is free from the commercial considerations that may influence the results of other search sites. To find queer resources, select Society and Culture, then Gay, Lesbian, and Bisexual or Transgendered. These sections lead to subsidiary directories arranged in terms of both groups (e.g., bisexuals, youth) and topics. The Sexuality section also includes links of interest to queers, with sections such as BDSM, polyamory, sacred sexuality, and sexual politics. The Open Directory Project also provides access to a large number of LGBTQ mailing lists and discussion forums.

Rainbow Query
http://www.rainbowquery.com

Rainbow Query specializes in helping the user "search the queer Internet." In addition to typical search functionality, the site also provides a directory structure arranged by both populations (gay men, youth, transgender, etc.) and topic areas (arts and entertainment, health and fitness, travel, money, etc.). Directory listings include both queer sites and mainstream sites (for example, the Arts and Entertainment section includes both Frameline and E! Online), with the queer-specific sites generally listed first.

Yahoo
http://www.yahoo.com

Yahoo is the best-known Web directory, and also features a search engine powered by Google. Because it is the oldest and most popular directory, many Web site creators submit their sites for inclusion in Yahoo's directory, and its sheer size can make it a bit cumbersome to navigate. For queer resources, select Society and Culture, then Cultures and Groups, then Lesbians, Gays and Bisexuals or Transgendered. Within the Lesbians, Gays, and Bisexuals category there is an entire subsidiary directory of interest to queers, with subsections such as arts and humanities, history, law, politics, and religion. There is also a listing of queer search engines and directories, and links to queer Usenet newsgroups.

NON-WEB RESOURCES

Although many users tend to think of "the Internet" and "the Web" as interchangeable, the Internet includes much more than just the World Wide Web. In fact, the Web was a relative latecomer to the online world, following electronic mailing lists, Usenet newsgroups, Internet Relay Chat (IRC), and several other resources—such as Bulletin Board Systems (BBSs) and Gopher—that are no longer widely used. Although the Web has come to dominate online information gathering, some of the older tried-and-true communication tools still exist and continue to be used for research and other purposes.

Electronic Mailing Lists

Electronic mailing lists are collections of e-mail addresses of people who are interested in sharing information about a specific topic. These range from small lists set up by groups of friends to much larger, established, public lists that can be accessed in a variety of ways.

To join a discussion on a mailing list, one must subscribe. Two common programs are used to manage mailing list subscriptions: Listserv and Majordomo. To subscribe to a list that uses one of these interfaces, send a request to the computer or "server" that hosts the list.

In addition to commands for subscribing and unsubscribing, both Listserv and Majordomo support many other functions that allow users to view archives of past messages, see the names of other subscribers, or receive mail as a single daily digest instead of individual messages. To get a list of commands and how to use them, send a message to "listserv" or "majordomo" at the appropriate server that contains the word "help" in the body of the message (not the subject header).

Basic Listserv and Majordomo Subscription Information

To subscribe to a Listserv mailing list, send a message to "listserv" at the server computer. For example, send e-mail to <listserv@brownvm.brown.edu> to subscribe to one of the Brown University lists. Leave the message subject header blank. In the body of the message enter:

Subscribe listname YourFirstName YourLastName

For example:

Subscribe bisexu-l Tekla Jones

Majordomo works similarly but requires your e-mail address instead of your name. Send e-mail to "majordomo" at the server computer of interest, for example, <majordomo@queernet.org>. It does not matter what, if anything, you put in the message subject header. In the body of the message enter:

Subscribe listname your e-mail@address

For example:

Subscribe gl-asb tekla@wahoo.com

Listerv or Majordomo will send you a message back either confirming your subscription request or giving you further instructions if your subscription attempt did not work. For more information on Listserv, see <http://www.lsoft.com/manuals/1.8d/user/user.html>. For more information on Majordomo, see <http://www.greatcircle.com/majordomo>.

Fortunately, Web interfaces have made mailing lists easier to use. Many e-mail lists now have Web sites that allow users to simply check boxes or buttons to subscribe, unsubscribe, or change from individual messages to a digest. An ever-increasing number of mailing lists are hosted at free sites such as Yahoo Groups rather than on personal or university servers. This saves list administrators a lot of work, but often comes at the price of small ads appended to messages. Using a Web interface, you can either sign up to have messages sent to you through e-mail or read messages directly from the Web.

Yahoo Groups

Yahoo Groups (http://groups.yahoo.com), formerly eGroups, is home to a large number of queer mailing lists. If you know the name of the list you want to join, type it into the search box. If not, you can browse through the directory: most queer groups fall within the Culture and Community/Groups section. From here, select Lesbian, Gay, and Bisexual or Transgendered. As of this writing, there were nearly 4,000 gay, lesbian, and bisexual mailing lists and over 400 transgender lists. Other alternative sexuality groups (e.g., BDSM lists) are categorized under Romance and Relationships/Adult. Many lists are open and you can join by simply clicking the "subscribe" link. Others are closed, and subscription requests are sent to a human list administrator for approval.

Many other LGBTQ mailing lists are hosted by QueerNet, which is described in Chapter 3, "Major LGBTQ Internet Research Tools."

Usenet Newsgroups

Usenet is a large hierarchy of topical discussions called "newsgroups" that follow a "bulletin board" model. Users post their messages in a public forum that any other user can read. One does not have to subscribe or become a member to post or read newsgroup messages. The free-wheeling nature of Usenet—while one of the defining features of the early Internet—has led to problems in the years since Usenet originated in the late 1970s, as millions of new users have signed on. One is the development of "flame wars," arguments that often devolve into personal insults and stray far from the intended topic. Another is "spam," the widespread proliferation of irrelevant, often commercial messages. Some newsgroups use moderators—humans who review each message for appropriateness—to reduce these problems. Many newsgroups have a "FAQ" (a document with answers to frequently asked questions) to avoid repetitive discussion of basic issues. For an archive of Usenet Frequently Asked Question, visit <http://faqs.org>. For general information about Usenet, see <http://www.westwords.com/guffey/netuse.html>.

Usenet can be accessed by a variety of newsreaders, including the Unix "rn" command, Free Agent for Windows, NewsXpress, Newswatcher for Mac, and the Netscape and Internet Explorer browsers. For information about and access to various newsreaders, see <http://www.newsreaders.com> or <http://www.newsguy.com>.

Once you're on, the trick is to find the discussion topics that interest you. Usenet is divided up into several top level sections. Most queer newsgroups fall within the **soc** (society) section. These include the heavily trafficked **soc.motss** (for "members of the same sex"), **soc.bi**, and **soc.support.transgendered** (interestingly, there is no **soc.lesbian**). The **talk** hierarchy includes discussion and debate about current topics. The **misc** section contains a broad range of topics, including health and legal issues. The **alt** hierarchy is "anything goes," and is the place for postings of an adult nature. There are also regional sections for different countries and different parts of the United States. Barnett and Sanlo in *The Lavender Web:LGBT Resources on the Internet* (http://www.lgbtcampus.org/resources/internet_chapter.html) provide a partial listing of LGBTQ newsgroups.

Google provides a Web interface (formerly DejaNews) that streamlines the Usenet reading and posting process. Users can search for specific keywords or phrases using a search engine that scans a Usenet archive. Specific searches work best. Simply doing a search on the keyword "gay," for example, returns over 200,000 matches, ranging from a message about gay bikers from **uk.rec.motorcycles** to a message about how to tell if someone is gay from **alt.gossip.celebrities**.

Deja has been archiving past Usenet postings since 1995, and the news archive now contains over 500 million messages.

Chat Forums

Chat forums allow users to converse in "real time." Users who are present in a chat room at the same time can send messages that appear instantly, and other users can reply. Chat forums have evolved their own language, replete with abbreviations and "emoticons" (such as the :-) smiley face) to allow rapid communication.

The granddaddy of chat forums in Internet Relay Chat, or IRC. Specific discussion forums are known as "channels," and are denoted by the # sign. Users can join one of several existing, ongoing chats, or can create private channels of their own. There are many IRC "clients," or interface programs, available, which will not be discussed here. For more information, see <http://www.irchelp.org>.

Many portal sites, such as Gay.com and Yahoo, have their own chat forums. A number of specialty sites also feature chats. Talk City (http://www.talkcity.com) is a Web site that provides access to a large number of chat forums, referred to as "neighborhoods." Most queer-oriented chats can be found in the Ethnic and Lifestyle section, under Alternative Lifestyles.

ICQ

ICQ is a program that lets you see who is logged on at the same time as you and to exchange instant messages and exchange files in real time. America Online (AOL) has a similar program called Instant Messenger. To use ICQ, download the program and register; you will be assigned a unique ICQ ID that is used to track when you are logged on. For information on ICQ and how to use it, see <http://www.icq.com/products/whatisicq.html>.

Other Tools

In addition to mailing lists, newsgroups, and chat, there are several other tools used to exchange information using the Internet. File Transfer Protocol (FTP) is a method of transferring files from one computer to another. You can use either a set of FTP commands or a client program that allows you to transfer files by dragging and dropping or clicking buttons. Many sites allow anonymous FTP, which lets you log on to a remote computer and download files without having a user account or password on that machine. Telnet is a remote log-in protocol that allows you to access other computers. Gopher, Archie, Veronica, and Wide Area Information Servers (WAIS) were among the earliest tools used to index and search the Internet, preceding modern search engines. For more information about a wide range of Internet tools—including e-mail, Usenet, and chat, as well as the more obscure programs—see Lesbian.org's *Introduction to the Internet* (http://www.lesbian.org/internet-guide/guide.html), the *WURD Internet Tutorial* (http://www.wurd.com/eng/ABCs/tutorial/index.htm), and *Learn the Net* (http://www.learnthenet.com).

CREDIBILITY

Many people use the Web to conduct research, but it is not always easy to determine the reliability of information in cyberspace. There are many reasons for this. Anybody can publish on the Web, and it is not filtered through the paper-based publishing world (with its peer review, and legal and professional requirements and norms). Additionally, the online world has not yet established a widely used means of verifying information. Online information is dynamic—a Web page can be up one day and then gone or radically altered the next, and some sites personalize information based on your behavior on the site. Dynamic information is, of course, more difficult to verify. Finally, commercialism currently pervades the Web, to the extent that informational content and advertisements may be difficult to distinguish.

If you are using the Web for research that you intend to publish, be aware that there are some cultural biases against digital information that derive from a long-standing fetish for the printed word. Print symbolizes authority, irreproachability, and permanence—due not only to the physical presence of a book or journal, but also to the perceived quality assured by print publishers' gatekeeping role. Digital information, on the other hand, is largely detached from the realm of intellectual and cultural authority.

So, while it is often difficult to determine the credibility of information on the Web, there are some indicators to look for and some questions to consider:

- Is there an author listed? Are you given any means to contact the author or the publisher of the information?
- If the information is offered by an organization, is it a reliable organization? Check to see if it is an organization that is regulated, approved, or accredited. Regulation and accreditation do not necessarily ensure accuracy or truth, but they do help to protect against claims of unreliability. For example, government and university sites (typically ending in .gov and .edu respectively) are more likely to have regulatory pressures.
- Does the site creator offer a mission statement, goals or objectives? Does the site's mission impact the credibility of the information?
- Is there a source given for data, or is data simply presented as factual?
- Is there a date given for the information? Is it current?
- If the information is the result of the a Web page author's personal research or experimentation, does the author describe a reliable methodology?
- If the author quotes other online sources, are the sources linked and available?
- Are other sites linked to this site? This can be an indicator of either the site's popularity or the degree to which the information is seen as valuable. One way to determine this is to check how the site is listed at <http://www.google.com>. Google is a search engine that factors in other sites' links to a particular site in ranking the relevance of search results. What do other sites say about this site?
- Does the site make clear distinctions between advertisements and other types of content? Beware if the information is presented in the context of an advertisement that is intended to persuade you to buy something.

A final word on protecting yourself: if you intend to reuse information from the Web, it is advisable to save or print the page for your records, because the site may change at any time. If your source is ever challenged, your saved or printed copy of the page will be the only proof of your source.

CITATIONS

Scholars have been struggling for well over a decade to create and maintain citation standards in cyberspace for the creation of footnotes. The scholastic community has not accepted, to date, a universal standard for citing Internet sources. In an attempt to solve this problem, professional organizations, such as the Modern Language Association (MLA) and the American Psychological Association (APA), are in the process of creating citation conventions that seek to demonstrate efficient and unambiguous references to Internet sources. Based on traditional approaches to scholarship, the emphasis continues to focus on refereed, authoritative sources as well as on historical texts, which historically have been and continue to be printed materials. As a result of this emphasis, scholars want citations of electronic sources to accomplish the same ends and—to the degree possible—have a format analogous to those of print sources. Creating citation conventions for Internet sources, nonetheless, is not a definitive process. Citation standards will doubtless change as technology, scholarly uses of electronic materials, and electronic publication practices evolve.

Crucial pieces of information for an Internet citation are the complete address (e.g., URL), identifying information (e.g., author's name, title); the date of publication, if possible; and the date of user access.

The variety of electronic publications include scholarly projects or information databases; professional or personal sites; online books; articles in online periodicals; publications on CD-ROM, diskette, or magnetic tape; works from an online service; and other electronic sources (such as sound clips, interviews, e-mail communications, and online postings). Categories of resources on the Internet also include FTP sites, telnet sites, Gopher sites, electronic mailing list messages, and Usenet postings.

To maintain accuracy when citing material from an electronic mailing list, newsgroup, chat room, or e-mail, the original contributor should be contacted and told about the intent to cite to make certain the information cited is as accurate as possible. Furthermore, the contribution should be quoted as fully as possible to ensure that the original context is preserved. Finally, it is recommend that you download and/or print Internet source pages, e-mail messages, etc., as an interim practice until the scholastic community accepts a universal standard for citing Internet sources. This practice will enable you to provide evidence of your sources should they change or disappear altogether, as often occurs on the Internet.

The following citation examples are in accordance with the guidelines in the *MLA Handbook for Writers of Research Papers,* Fifth Edition, by Joseph Gibaldi, The Modern Language Association of America, New York, 1999. The MLA standards are used in the liberal arts and humanities and in most of the theoretical works in queer studies. However, if you are conducting research in the social sciences, you will—most likely—use the APA standards (a URL for those standards is listed later in this section). Also, the citations in this guide follow MLA standards except in Chapter 9, "Social and Biological Sciences," in which we use APA style, as is common to those disciplines.

Examples of Citation Styles

Professional or Personal Site

> *CLAGS: Center for Lesbian and Gay Studies, City University of New York (CUNY).* 18 May 2000. CLAGS, City University of New York. 23 Aug. 2000. <http://web.gsuc.cuny.edu/clags/>.

First, list and italicize the Web site title. Next, record the date of creation of the Web site. List the publisher of the site, then the date of user access. Finally, include the URL.

Article in a Scholarly Journal

> Swartz, L.H. Ph.D. LL.M., R.N. "Legal Implications of the New Ferment Concerning Transsexualism." *The International Journal of Transgenderism* (IJT) 2/4 (1998). 23 Aug. 2000. <http://www.symposion.com/ijt/ijtc0604.htm>.

List the author, with the last name first. Next, list the title of the article, in quotations. List and italicize the journal title next, with the volume or edition, followed by the publication date. List the date of user access afterward. Finally, enter the URL.

Article in a Newspaper or on a Newswire

> Oxenberg, Christina. "Hellfire and Khakis." *Salon.com*. 23 Aug. 2000. 23 Aug. 2000. <http://www.salon.com/sex/feature/2000/08/23/hellfire/>.

Similar to the example listed in 2.

E-Mail Communication

> Rubin, Gayle S. "Re: National Organization for Women SM Policy Reform Project." E-mail to Sanda Steinbauer. 16 May 1999.

List the name of the e-mail sender, with the last name first. List the subject of the e-mail in quotations. Record the name of the recipient. Finally, list the date the e-mail was sent.

Online Posting

> Hall, Lesley. "Usual invitation to introduce oneself." Online posting. 27 Mar. 2000. Histsex: For Historians of Sexuality. 27 Mar. 2000. <http://histsex.listbot.com/>.

List the author of the posting that you are citing, last name first. Next, list the title of the author's posting in quotations. Record that it is an online posting. List the date of the posting, where it was posted (for example, on a Web site or an electronic mailing list), and the date the post was received. Finally, if available, list the URL where this posting exists on the Internet.

Refer to the following sources for more complete guidelines and models for documenting Internet sources. Guidelines are certain to change over time, and these sources will have more up-to-date information.

> Walker, Janice. *MLA-Style Citations of Electronic Sources*. Vers. 1.0, 20 Mar. 2000. Columbia University Press. 23 Aug. 2000. <http://www.columbia.edu/cu/cup/cgos/idx_basic.html>.

The previous resource is endorsed by the Alliance for Computers and Writing (ACW). The ACW is a national, nonprofit organization committed to supporting teachers at all levels of instruction in their intelligent, theory-based use of computers in writing instruction. The ACW Web site is <http://english.ttu.edu/acw/>.

> Harnack, Andrew and Kleppinger, Gene. *Beyond the MLA Handbook: Documenting Sources on the Internet*. 10 Jun. 1996. Eastern Kentucky University. 23 Aug. 2000. <http://english.ttu.edu/kairos/1.2/inbox/mla_archive.html>.

This is an excellent resource, especially for addressing issues not covered by the *MLA Handbook*.

> American Psychological Association. *Electronic Reference Formats Recommended by the American Psychological Association*. 22 Aug. 2000. American Psychological Association. 23 Aug. 2000. <http://www.apa.org/journals/webref.html>.

Note: See *Copyright and Permissions* <http://www.apa.org/journals/copyrite.html> for policy on distribution and reuse. 19 Nov. 1999. This document replaces *How to Cite Information From the Internet and the World Wide Web.*

> Columbia University Press. *Basic CGOS Style*. 20 Mar. 2000. Columbia University Press. 23 Aug. 2000. <http://www.columbia.edu/cu/cup/cgos/idx_basic.html>.

For more information and examples, see Walker, Janice R. and Taylor, Todd. (1998). *The Columbia Guide to Online Style*. New York: Columbia University Press.

> International Federation of Library Associations and Institutions. *Library and Information Science: Citation Guides for Electronic Documents*. 30 Sept. 1999. International Federation of Library Associations and Institutions. 23 Aug. 2000. <http://www.ifla.org/I/training/citation/citing.htm>.

This resource provides excellent links to style guides and other resources on how to cite Internet sources.

Li, Xia and Crane, Nancy B. (1996) *Electronic Styles: A Handbook to Citing Electronic Information* (Revised Edition). Medford, NJ: Information Today.

Refer to this guide for more complete recommendations on bibliographic formats for electronic sources. The book follows two common citation conventions, APA and MLA, and embellishes with the unique features of electronic information.

Li, Xia and Crane, Nancy B. (1993) *Electronic Style: A Guide to Citing Electronic Information.* Westport, CT: Mecklermedia.

Many scholars currently follow this reference standard.

International Organization for Standardization (ISO). Excerpts from ISO Draft International Standard 690-2 - Information and documentation—Bibliographic references—Electronic documents or parts thereof. 21 Mar. 2000. National Library of Canada. 23 Aug. 2000. <http://www.nlc-bnc.ca/iso/tc46sc9/standard/690-2e.htm>.

Hale, Constance, Scanlon, Jessie, and Scanlon, Hale (1999). *Wired Style: Prinicples of English Usage in the Digital Age.* New York: Broadway Books.

A general style guide produced by the publishers of *Wired* magazine. Good source for proper appearance of URLs, spelling of technical jargon, etc. Includes a short section on citations.

Chapter 3

Major LGBTQ Internet Research Tools

Alan Ellis
Liz Highleyman

There are a number of excellent LGBTQ Internet research tools available, including the two major LGBTQ portals, several additional portals and online directories, an online database of more than 8 million journal articles, and several non-Web resources. Regardless of the area you are interested in, these sites offer superb resources for Internet research. You can use Gay.com, PlanetOut, and the other sites listed as follows to find hundreds of links to queer-related sites to help you determine a research topic and to find information and reference materials once you have chosen your topic.

MAJOR LGBTQ PORTALS

The two major portals for the LGBTQ communities are Gay.com (http://www.gay.com) and PlanetOut (http://www.planetout.com). Both sites offer current news, chat rooms, discussion boards, event calendars, search functions, and links to hundreds of community-related sites. On November 16, 2000, PlanetOut and Gay.com announced an agreement to merge. The merged company intends to maintain both sites, and we describe both as if they were independent. The merger of the two largest Internet businesses serving the LGBT market will create a global media and services company that reaches more than 3.5 million unique individuals a month and counts more than 1.6 million registered users.[1]

Some have expressed concern about the merger of the two largest LGBTQ Web properties under a single parent corporation. Critics contend that the merger could create a near-monopoly that will focus on market considerations and narrow the scope of available LGBTQ news and information; they also fear that the diversity of voices and viewpoints will decrease. According to longtime queer activist Bill Dobbs, "As a minority community, we're already vulnerable to our media outlets and who runs them. This concentration of ownership and loss of competition makes us even more vulnerable."

Gay.com
http://www.gay.com

Gay.com describes itself as the "largest and most extensive online network targeted to the lesbian and gay community." Gay.com also has resources and links for bisexuals and transgendered individuals. As of November 2000, Gay.com had 850,000 registered users. The site offers news, search, shopping, chats, message boards, and other online services (including online commerce and marketing programs that focus on entertainment, finance, health care, telecommunications, and travel services). Gay.com also offers several international sites. When you visit Gay.com, you will be asked to select which of these sites you wish to visit, including Gay.com UK, Gay.com Italia, and Gay.com Francophone. Gay.com is based in San Francisco, with offices in New York, London, Paris, and Buenos Aires.

PlanetOut
http://www.planetout.com

PlanetOut was established in 1995 and offers original content including news, entertainment, travel, finance, career information, an Internet radio channel, personal ads, free e-mail, community centers, shopping, message boards, chats, and advice columns by writers such as Betty DeGeneres. As of November 2000, PlanetOut had more than 850,000 registered users (the same number as Gay.com). PlanetOut is also based in San Francisco.

Note: At the end of this chapter, you will find interviews with Mark Elderkin, cofounder of Gay.com and Rhona Berenstein, Vice President of Strategic Marketing at Gay.com; and Megan Smith, former Chief Executive Officer (CEO) of PlanetOut.com, who will become the new president of the merged organization.

ONLINE DIRECTORIES AND OTHER WEB SITES

In addition to the two major LGBTQ portals, there are a number of other sites that provide online information helpful to researchers. About.com, Homorama, Lesbian.com, LesbiaNation, and the Queer Resources Directory (QRD) are among the best of these sites.

About's Gay/Lesbian Issues Site
http://gaylesissues.about.com/newsissues/gaylesissues/inex.htm

The About network consists of over 700 sites that cover more than 50,000 subjects and over 1 million links. About describes itself as "the fastest-growing archive of high quality original content." About was previously known as the Mining Company, which was formed in 1997 and renamed in 1999. In the area of LGBTQ studies, the Gay/Lesbian Issues section provides links arranged by various categories including hate crimes, U.S. government resources, history, politics and government, religion, and workplace issues. Although the section's title is limited to gays and lesbians, it also includes links dealing with bisexuality and transgender concerns. Current political issues in the queer community can be identified here. About also has sections dealing with other alternative sexuality issues, such as BDSM and polyamory, but access to this and other material deemed "adult" requires you to register with AdultCheck (http://www.adultcheck.com), an age verification system (to register, you must be twenty-one or older; registration costs about $20 per year).

GayWired
http://www.gaywired.com

After Gay.com and PlanetOut, GayWired is one of the largest and most extensive gay sites on the Web. Like Gay.com and PlanetOut, it is a complete portal featuring several channels. Activism alerts, "Religious Right Watch," press releases from LGBTQ organizations, and columns by writers such as Rex Wockner, Patricia Nell Warren, and Mubarak Dahir can be found in the News section. The Life section includes information related to health, HIV, transgender issues, women, families, and youth. The Play section includes listings of LGBTQ events such as Pride celebrations and film festivals. The Guide section includes information about education, history (including several good feature articles), marriage issues, political and legal issues (including news from Lambda Legal Defense and Education Fund), and religion and spirituality. The site also features entertainment and travel information, and local sections with news and events for many U.S. and international cities.

Homorama
http://www.homorama.com/homorama/index.html

Homorama provides a good search function and links categorized by such topics as education, hate groups, history, law and politics, personalities, workplace and youth. It also provides links to less known or less-often-accessed Web sites that provide alternative viewpoints and information.

Independent Gay Forum
http://indegayforum.org/

The Independent Gay Forum was created by a group of gay writers, academics, attorneys, and activists "who feel dissatisfied with the current level of discussion of gay-related issues." Paul Varnell is the editor of the site, which includes articles by authors such as Bruce Bawer, Andrew Sullivan, Stephen O. Murray, Eric Marcus, and Jonathan Rauch. Titles of articles include "Five Reasons I Don't Take Queer Theory Seriously" (S. Murray) and "Gay People Want to Get Married, Too" (E. Marcus). Other articles tend toward a more moderate or conservative gay (mostly gay male) perspective.

Lesbian.com
http://www.lesbian.com

Lesbian.com is one of the most comprehensive portals for lesbians. In keeping with lesbian feminist politics, there is substantial content on activism, politics, antioppression, and specific identity groups (e.g., deaf lesbians, elders/crones, fat dykes, lesbians of color). The site has a clean look and is much less commercial than the major gay portals. It is set up in a familiar directory format with more than a dozen top-level categories. As with most portals, there are sections on business, family, health, and travel, but Lesbian.com also features several unique sections such as gender (including butch and femme content), herstory, and "land dykes" (for women who live on or are interested in lesbian land).

Lesbian.org
http://www.lesbian.org

Lesbian.org is a simple site with a mission of "promoting lesbian visibility on the Internet." The list of links is rather small compared to a site like Lesbian.com, but includes some gems such as an annotated bibliography on butch/femme issues and Lesbian.org's *Introduction to the Internet,* which explains Internet tools, resources for new users, netiquette and more. Lesbian.org is the home of several mailing lists for queer women, including **poly-dykes, travel-dykes,** and **lesbian-writers.** These lists are managed using Majordomo; instructions for joining can be found at <www.lesbian.org/lists.html>.

LesbiaNation
http://www.lesbianation.com

LesbiaNation is a more typical commercially-oriented portal, with channels for entertainment, sports, health and wellness, travel, news, and shopping. Each section contains a selection of original articles, including columns such as Felice Newman's sexuality series. Access to some of the content requires free registration. The site also includes chat and discussion areas.

Queer Resources Directory
http://www.qrd.org

The Queer Resources Directory (QRD), which describes itself as "an electronic research library specifically dedicated to sexual minorities," may well be the most useful site on the Internet for researchers studying LGBTQ issues. Most links on the site are current, and information and links are updated on a fairly regular basis. QRD is also one of the oldest Web sites, predating most of the commercial Web sites such as Gay.com and PlanetOut. Ron Buckmire started QRD in 1991 as an electronic archive for the activist group Queer Nation; it soon expanded into a broader directory covering a wide range of topics of interest to the LGBTQ community. The site is run by volunteers using donated equipment. As of January 2001, the site included more than 25,000 files arranged in the following categories: queers and their families, including parenting and marriage; queer youth; queers and religion, and religious queers; queer health, including safer sex information; electronic resources; queer media, including magazines, television, movies, and more; queer events, including conferences and celebrations worldwide; queer culture, history and origins; worldwide queer information from Australia to Zimbabwe; business, legal, and workplace issues, including information on domestic partnerships and queer-friendly businesses; politics, political news, and activism; and organizations, directories, and newsletters. QRD's goal is "to contain every scrap of knowledge which has been used in or is part of the struggle for full equality."

The previous section describes only a sample of the Internet's many LGBTQ resources. Other sites, including Bgay (http://www.bgay.com/), Gay Crawler (http://www.gaycrawler.com/), and GayScape (http://www.jwpublishing.com/gayscape/classic.html), also feature collections of links to LGBTQ resources.

ONLINE DATABASE OF JOURNAL ARTICLES

Ingenta
http://www.ingenta.com

Ingenta provides free access to article summaries from over 20,000 publications. You must be a subscriber or pay per article to view the full text. The database can be used to identify articles in journals that can be accessed in most large university libraries. This is an excellent resource for identifying articles on a specific LGBTQ research topic in most disciplines and areas of academic study. Entering the term "gay" returns over 2,000 articles ranging from popular media to academic journals. Ingenta acquired the UnCover article database in 2000.

MAJOR LGBTQ NON-WEB RESOURCES

Lesbian Mailing Lists
http://www.lesbian.org/lesbian-lists/

Finnish lesbian Eva Isaksson has put together a comprehensive collection of mailing lists of interest to lesbians (several explicitly include bi women as well). Here you can find discussion lists for everyone from Asian/Pacific Islander lesbians and bi women, to lesbian academics, to members of the Lesbian Avengers—even the Gillian

Anderson Estrogen Brigade! Each list name on the site is a live link leading to information about how to join the specific list.

LGBT+ Internet Mailing Lists
http://www.qrd.org/electronic/email/

The Queer Resources Directory hosts an extensive "list of lists" featuring nearly 400 queer mailing lists. In addition to mailing lists for gay, lesbian, bisexual, and transgender communities, there are also discussion groups covering various sexual minorities and alternative sexualities, families and allies of queers, and more. Each mailing list has a live link leading to information about the list's purpose and how to join.

QueerNet
http://www.queernet.org

QueerNet, founded by Roger Klorese in 1991, is home to many of the Internet's most popular gay, lesbian, bisexual, transgender, and leather/BDSM mailing lists. You can use the QueerNet Web site to browse an index of over 500 existing mailing lists, organized into categories including activism, community organizations and politics; geographic, ethnic, and religious/spiritual; interests, professions and pastimes; AIDS and HIV; sex and fetish; and a catch-all section for social, support, discussion, and information. Many mailing lists in the index have live links to their own sites. You can request to join QueerNet lists simply by clicking a few buttons; many lists are open to the public, although some require approval by a list owner. You can also set up and manage your own mailing lists. On January 1, 2001, QueerNet became a project of the Online Policy Group (http://www.onlinepolicy.org), an organization that promotes equal access, privacy, and civil liberties in cyberspace. An interview with Roger Klorese and Will Doherty (executive director of the Online Policy Group) can be found at the end of this chapter.

Sappho
sappho-request@sappho.org

The Sappho mailing list—the mother of lesbian mailing lists—is quite possibly the longest-running and most well-known Internet discussion forum for queer women. It was started in 1987 at MIT by

Jean "Ambar" Diaz. The list is open to all women and has several hundred members. Any topic of interest to lesbian and bi women is fair game. Over the years Sappho has given rise to several local versions of the list, such as **ba-sappho** (San Francisco Bay Area), **boston-sappho,** and **euro-sappho.** To join the mother Sappho list, send e-mail to <sappho-request@sappho.org>.

Soc.motss

While there are many Usenet Newsgroups devoted to the queer community, perhaps the longest-running and best known is **soc.motss** ("motss" means "members of the same sex"). Like many popular newsgroups, **soc.motss** is often awash in flame wars, commercial postings, and irrelevant content, but it remains one of the world's largest and most wide-ranging online LGBTQ discussion forums.

Soc.women.lesbian-and-bi

This newsgroup is one of the main Usenet discussion forums for lesbian and bisexual women.

INTERVIEW: MARK ELDERKIN AND RHONA BERENSTEIN

Mark Elderkin is cofounder of Gay.com and President and Chief Operating Officer of Online Partners. Prior to staring Gay.com, Mr. Elderkin founded Volano, a Java-based software company specializing in Internet chat platforms. Elderkin has served as product manager and marketing director for National Semiconductor, Network Equipment Technologies, CellNet Data Systems, RadioMail Corporation, and Sprint PCS. He holds an MBA from Berkeley Haas School of Business and a BS in systems engineering from Boston University, where he graduated magna cum laude. Elderkin serves on the board of the Gay and Lesbian Alliance Against Defamation (GLAAD).

Rhona Berenstein, PhD, is Vice President of Client Services and Business Development at Gay.com/Online Partners. Prior to joining Online Partners, Berenstein was a consultant in online multimedia and educational technology. She is a specialist in gay and lesbian media and holds an MA and PhD from the University of California-Los Angeles. Her most recent academic post was Associate Professor and

Director of Film Studies at the University of California, Irvine. Berenstein serves on the National Research Advisory Board of GLAAD and is a member of the Board of Directors of the Los Angeles Gay and Lesbian Center.

HMI*: What can you tell us about the history of Gay.com?

Mark Elderkin: I acquired the domain name Gay.com in 1994 with my partner Jeff Bennett. We did not have a plan for it at that time, as it was prior to the release of the Netscape browser and before people had Web sites. People were using e-mail, so we decided to create a gay and lesbian newsletter telling people about what was happening worldwide. We started the subscription service in 1994 while we were employed by other companies. So we were doing it as a hobby. As we learned more about the gay community and technology, and with the later explosion of the Web, we built Gay.com in August 1996. We launched before the other gay Web sites and we started from the beginning as a community site, where we were connecting people via the Web.

HMI: Was it intended to be a chat room or a dating service?

Mark Elderkin: It was not a dating service, it was a means for people to connect, just like America Online's (AOL) chat room service. We really followed AOL's model, but without restrictions on what people could and could not say. At that time, we started to build an affiliate network so that we could offer our service to any Webmaster in any country. So, we started as a global site from the beginning. That was the start—1996 to 1997—and we focused on improving the technology to connect as many people as we could in one place at one time. Today we have numbers similar to Excite with 15,000 to 20,000 users chatting simultaneously. We recently merged with Gay.net—a consolidator of gay and lesbian media—and are now affiliated with major magazines and gay publications around the country, including San Francisco's *Bay Area Reporter* and the *Washington* (DC) *Blade*.

---

*Interview conducted on July 24, 2000, by Moe Sidaoui and Alan Ellis of the Board of Directors of HMI.

HMI: At the time you were first developing the Gay.com site, you had a separate business that was developing the Volano chat software—software that is now used by hundreds of Fortune 500 companies. What role did Volano play in the development of Gay.com?

Mark Elderkin: We actually used the Gay.com site as a test bed for the software. Users on Gay.com had a lot of good ideas about how to improve the software and about what they wanted, so it was a great place to test and develop chat software.

HMI: What is your relationship with other Web sites?

Mark Elderkin: We recently acquired OnQ, a site that has a great history [OnQ was AOL's gay channel]. OnQ made AOL use the word "gay." We also partner with many other Web sites.

HMI: What are the key services and features that Gay.com offers researchers in the queer community?

Mark Elderkin: Gay.com provides a variety of information that you cannot find anywhere else: headline news, information on current events, and other resources. We have writers around the world. So, a researcher can get information about issues of interest today, here and elsewhere. We also have message boards where a user can post a question or idea and get people's responses to it.

We have sections on arts, theater, movies, and music. We gather research from different pharmaceutical companies and publish information about their products and services in relation to HIV and AIDS. We also have sections about food and wine, fitness and health, and travel—what travel spots are hot.

HMI: If you were doing research on an academic topic, how might you use Gay.com?

Rhona Berenstein: If I needed to do extensive or in-depth research, I would go to the University of California at Berkeley's library or another major university library. However, researchers can use Gay.com to get a good sense of what is happening in the gay and lesbian world. In fact, we just launched our Latin American site, which is something no other gay portal has done. Especially for undergraduate research,

Gay.com can be a great resource. For graduate and postgraduate research, it can be a good starting place.

HMI: Can anyone post a question for his or her research? What does it take to become a member of Gay.com?

Mark Elderkin: Absolutely, anyone can post questions as long as they are not trying to solicit anything. Becoming a Gay.com member is free, and anyone can become a member.

HMI: What makes Gay.com unique to the LGBTQ community and, in particular to the LGBTQ academic world?

Mark Elderkin: Gay.com is unique in that it has information about what is happening around the world. We offer our information in seven different languages. Gay.com is among the top twenty-five Web sites—of all Web sites, not just gay Web sites—in numbers of return visits, page views, and time spent on the site. We have over 10 million visits each month, with more than 100 million page views. We are rated number one among gay sites.

Rhona Berenstein: Also, we are developing a huge database of information that includes speeches by politicians and other historical information. As this database grows, it will become an important resource for researchers and others interested in the history of our community.

HMI: Are you planning to go public?

Mark Elderkin: It would be nice to do so, but we are doing very well right now. We do not have to go public to raise cash like many other dot-com companies. We just raised 23 million dollars to further develop our site.

HMI: Do you foresee profitability?

Mark Elderkin: Yes, but growth costs money and right now we're focusing on growing globally, which requires opening more offices. We just opened offices in London, Paris, and Buenos Aires. We are establishing the dominant position in the market.

INTERVIEW: MEGAN SMITH

Megan J. Smith has been CEO of PlanetOut since 1998. Previously, she held the positions of president, COO, and director of operations at PlanetOut. Before she came to PlanetOut, Megan worked at General Magic for six years, serving as manager, European alliances; manager, partners and licensing; lead mechanical engineer for Magic Cap 1.0; and hardware team projects manager. Previously, she worked at Apple Computer Japan where she was responsible for multimedia market development. Megan holds a BS and MS in mechanical engineering from the Massachusetts Institute of Technology (MIT), where she completed her graduate studies in the MIT Media Lab. After graduation, Megan served on MIT's board of trustees for five years. In August 1999, she received the first Gay and Lesbian Alliance Against Defamation (GLAAD) Internet Leadership Award.

HMI*: How does the LGBTQ academic world benefit from PlanetOut?

Megan Smith: PlanetOut and our similar area on AOL [AOL keyword: PlanetOut] are destination Web centers, or portals. PlanetOut is an interactive media company, and these places online have a lot of value for academics both in terms of community and especially in terms of resources and content. For the researcher approaching a search from an activism or a news perspective, we've written five articles a day every weekday since 1996, in addition to having a partnership with *The Advocate* (http://www.advocate.com). So in that sense, a researcher would find original content. A well-known gay anthology recently asked us for all our headline news. We license it to them so that it can be on their CD, in their print materials, and on their Web site. In an academic context, we enable researchers to examine the last five years of gay and lesbian news. PlanetOut will be five years old this summer.

In addition, we help in a "card catalog" role in making sense of who's out there and what's out there. We host a tremendous amount of community dialog, and we also have a lot of deep databases for sourcing. I attended a lecture about pop culture during the San Francisco International Lesbian and Gay Film Festival, and the two

---

*Interview conducted on August 1, 2000, by Elizabeth Taylor for HMI.

sources that the lecturer used were PlanetOut and *The Advocate*. It was nice to see that people are really starting to see what's there and dive in.

We are working to make sure that PlanetOut is both a mile wide and a mile deep. Many people talk about many topics, but academics really need depth of information. Whether it's deep databases or extensive access to community people talking about topics, academics need an environment where people want to talk about the issues.

We also have a list of every known LGBTQ film festival in the world and a database compiled by Jenni Olsen, PlanetOut's Popcorn Q producer, that contains a comprehensive collection of gay and lesbian films [see description of Popcorn Q as follows]. Jenni comes from academia. She is a film archivist from the University of Minnesota who started the film festival in Minneapolis, then went to Frameline in San Francisco and codirected the San Francisco International Lesbian and Gay Film Festival for three years. She's an incredible person. Tom Rielly, PlanetOut's founder, considers her the ultimate person to do this work. One of her films, *Homo Promo,* is a chronological trailer of gay-centric films through forty years of film archives.

HMI: My guess is that academics in film and cultural studies are constantly accessing Popcorn Q.

Megan Smith: Yes. Tom Rielly had a really broad vision, and Popcorn Q is a great example of this. I think that you can also look at other channels. Tom sensed that it was best to bring the elite media creators together with the mass market. You want to make sure you are serving not only professionals and academics (for example, one area of Popcorn Q specializes in people who are in the business of film and film presentation), but also all of the consumers who are surfing or talking to each other or rating it or buying it or watching it. (We've had an online cinema since 1997.) What's nice about Popcorn Q is that it's a place where those two groups can meet. There's something there for both of them. Having the producers together with the consumers is really powerful. And the consumer also gets to produce something, whether just by stating an opinion (posting a vote on a poll) or by actually going out and making a film and entering it as an amateur.

The other big thing is the access we offer to other academics and the LGBTQ community. Academics who want to speak with people within the gay and lesbian community can talk directly to them by coming to the site and creating places to talk with other academics, have dialogs with community members, ask questions, and try to get some information.

Of course there are message boards and chat rooms on tons of different topics, plus free e-mail. There's lots of space for community sharing, so academic folks can pose a question to other people; they can create message boards or talk on existing boards with others.

As an interactive media company, we also run gay and lesbian areas on many of the predominant mainstream Internet portals. In other words, if a person is searching for gay material using Yahoo, they are accessing PlanetOut resources, because our resources are integrated into Yahoo. We are a partner of AOL, Netscape, and CompuServe, We also provide the gay and lesbian areas of online radio such as Real Networks. We also do print. We partnered with *The Advocate* and *OUT* magazines. We produce a television show with City TV and broadcast it on the 'Net.

So, from an academic perspective, we offer many different resources: tons of information about entertainment and community through our Web site, our AOL site, and through the mainstream portals. In addition, our news team is always looking for two to three sources on any given story, and they primarily do Web journalism. Their news is breaking twenty-four hours a day globally. They are using major news agencies as well as scanning wire services and sending e-mail to people they know to verify things. They're also looking at all the electronic mailing lists in gay space.

HMI: Does PlanetOut work closely with any academic organizations right now?

Megan Smith: We've talked with The New School a little bit. We think that it would be incredible to have online continuing education in gay and lesbian studies. We have a history channel that many people go to, which we produce with a partner, the GLBT Historical Society of Northern California. Fundamentally, they're very much an academic, historical research group. Even though it refers to Northern California in its title, it's actually much broader geographically.

HMI: Does PlanetOut currently offer or have any plans to offer key phrases for academics who are using your site?

Megan Smith: Yes. We have a partnership with Google on our search feature. If you are researching and consistently wanting to monitor a particular set of topics, you can go under keyword "search" and bring up links that are relevant to that word or phrase.

HMI: Does PlanetOut work with several search engines, such as Rainbow Query?

Megan Smith: No, we don't work with them because we find that Google is able to do what we wanted to do. We don't want to have several different searches—just one streamlined search, and Google is phenomenal. The mathematics behind it are the best. We looked at all the different suppliers, and it's definitely Google, hands down. From an academic perspective, it's wonderful to have PlanetOut searches layered with Google searches. What I love about this feature on the PlanetOut site is that you can save searches. You can save a search page and surf back to it. It's like saving your own permanent research page. You don't have to start over each time you want to explore that topic again; you can just return to the search page you saved.

HMI: How have your personal affiliations with academic organizations informed the ways that PlanetOut is structured?

Megan Smith: At MIT, I was on the board for four years, and I still serve on committees: the Visiting Committee for Engineering Systems, Athletics, and the Media Lab. One of the great strengths of academic organizations is the concept of collegiality. It's a less hierarchical style. Sometimes it can make an organization less efficient, but ultimately I believe that it is preferable to have less of that military perspective of hierarchy and who's the boss. So, seeing how groups work together, this influences a couple of key things at PlanetOut. We have a more collegial style. The second thing is that our community is run by a group of volunteers and different types of staff who are interested in particular topics. It's a relatively loose-knit volunteer network. So I think there is a tremendous amount of collegiality there, a more academic style, sort of electing to run something or create something. When someone says he or she is interested in whatever topic, they own the topic. Just like when a professor says, "I'm going to take my research this way."

HMI: What do you think makes PlanetOut unique to the LGBTQ community, in particular to the academic world?

Megan Smith: I think that PlanetOut is unique because of the interactive media mixture. We've got a large number of professional voices bringing media content—including news, travel, film, money, and career information. Within each of these channels we have databases of information that we're either licensing or creating. We also have experts. For example, we have an incredible career expert who talks about gay and lesbian career issues. We have a great personal finance expert who talks about money issues, especially for couples. We have experts on legal issues. We have Betty DeGeneres (Ellen's mom), who is probably our most famous columnist, dealing with all kinds of coming out and acceptance issues. She's got everybody from gay and lesbian people to parents and friends and family and others circling around that universe. We've got different fun, extreme voices on the topics of music and entertainment. On top of this, there's a higher volume of interactive media created by our members and visitors. So on the media side, there is a huge diversity and wealth of deep information on many different topics that lots of people can access. This information wasn't available or distributable before PlanetOut. We changed all of that, because we can bring this wealth of information and combine it with interactivity. When people visit our site they can start a dialogue or chat.

HMI: Is it possible simply to send an e-mail to an expert and get a response?

Megan Smith: Yes. We have about twenty topic experts, as well as category experts, and our regular volunteer feedback crew. We get quite a range of e-mail. We get about two percent hate mail, but many more people use us as a resource.

HMI: What would you say your rate of response is for incoming e-mail? I know that you expected 900,000 PlanetOut members by December 2000, so that means a lot of e-mail.

Megan Smith: We have a whole crew working on it, so for the most part people get answers relatively quickly.

HMI: You did your undergraduate work and your master's degree at MIT, both in mechanical engineering. But your affiliation for your master's was in the Media Lab. Did you do any research from the LGBTQ perspective?

Megan Smith: My academic research for my master's was related to kids, gay or straight. It related to "usability," focusing on a particular consumer group or a particular group within the population. It was focused on how you could feel what you were seeing on a computer screen; it was an electromechanical project. Since kids don't tend to do well with typing, I was exploring other ways for kids to interact with a virtual space—for example, by feeling a three-dimensional environment. There are many applications out there for things like tactile simulation and remote environments.

I also did another project right after that one. I won some money—it was called the Carol Wilson award—and I went down to South America to severely deforested areas where people have no cooking fuel to understand more about the problem. That problem has existed for a long time and is getting worse. Unfortunately, the people who make solar and alternative cooking stoves often produce stoves that have no relevance to the people who need them. One of the fundamental things about a culture is its food. You can't go in and ask people to suddenly cook their food for four hours and have a soup if they mainly eat hot, fried foods. So what I was doing in South America was not so much designing a stove, but rather talking to women all over Ecuador and Bolivia about design criteria for these products.

Ecuador has twenty-five of the world's twenty-eight climactic zones, so you get these extremes of how cooking has been solved. Examples range from people in the coastal regions who cook their food way out in the backyard because the climate was very hot, to people in the inner mountains who had the fire in their house and didn't care that it was smoking them out because it kept them warm, to people in the jungle who had their cooking things outside but down on the ground for morning warmth. There were a lot of different cultural reasons for the differences, including the temperature issue. I worked with Peace Corps volunteers to get access to the people and talk to them. What was so fundamental about that work—and what I love about the Internet also—is that you're asking people in a much more respectful way what they are doing, what they need, what's the design criteria for this thing. For the stove project, I tried not to sit up in Boston and design something for people I had never met. In the same way, what's so nice about the 'Net is that you have instant dialog and instant feedback from customers and participants.

HMI: Before PlanetOut, what do you remember being available on the 'Net in relation to LGBTQ resources?

Megan Smith: There were many things actually. But the two biggest Internet things were Digital Queers and the Queer Resources Directory (QRD). Tom Rielly founded Digital Queers in 1992, and helped a lot of the dot-org sites get themselves to really move. Having grown up professionally in Silicon Valley as a media guy and a marketing technician, and having all these computers and all this access, when the AIDS crisis hit and he became involved with the nonprofits, he was just stunned by the total lack of that kind of access for people with AIDS and the organizations that supported them. That's why he founded Digital Queers. His contribution to the digital space is massive; in those early times what he foresaw was a tool for everybody to get together and get information. Ultimately, Tom envisioned not only that early communication, but also broadband digital media and the ability to get *The Advocate* and other magazines properly distributed. Magazines, like *Out* and *The Advocate*, that only have a circulation of 100,000 or 88,000—it's ridiculous. Many more people than that would like to read them. He envisioned this ability to get there. So really, PlanetOut comes out of that early vision.

What happened on the 'Net at that time is that Digital Queers members would come together and have parties, and he would lobby all his buddies to donate what they could based upon their specialization. This way he could amass an incredible set of equipment. Then he'd have what he called "Beauty Makeovers." The dot-orgs would have half a broken computer and a laptop. He'd ask the employees of the dot-orgs, if they could have anything they wanted what would it be. He encouraged them to always think bigger. "Would you all like a laptop? A network?" In this way, he would make over groups like Gay and Lesbian Alliance Against Defamation (GLAAD), the National Gay and Lesbian Task Force (NGLTF), and Parents and Friends of Lesbians and Gays (PFLAG), etc. Each year Digital Queers would target a group and convince the group to hire an information systems person, make sure there was enough money to do that, and then hire people to install the computer and network systems.

Tom also befriended an incredible guy named Ron Buckmire (who was a very early PlanetOut board member) who started QRD in 1991. It was all volunteer-based. This was the first time that there was a surfable, topical set of LGBTQ information. People submitted documents into folders and Web-enabled them so that users could surf around and find all kinds of information. I remember there was a category called "People That Hate Us," where you could go and find all of Pat Robertson's *700 Club* issues. In the early days, PlanetOut tried to partner with QRD and bring that research in and make it much more widely available and searchable, but we ended up not being able to do it because there was so much non-copyrightable material. For example, people had taken stuff from corporate memos from executives explaining why they couldn't have domestic partner benefits. Amazing things. Buckmire is a mathematician at Occidental College in Los Angeles and QRD comes straight out of the university world of the early Internet, when the Internet had some great stuff. So QRD was almost like a queer Library of Congress. All of the dot-orgs were coming online at the time, as well as Digital Queers, and Tom was also pushing AOL sign ups, and working with QRD to get people online and on the Web.

One of the other big early hubs was created by David Stazer who was a student at the University of California at Berkeley. He's now Executive Vice President of Product Development at PlanetOut. David founded a site called Queer Info Server on the Web, which was kind of like a gay Yahoo. David was archiving all gay- and lesbian-centric sites he could find on any topic. He was also doing an amazing thing called People Out on the 'Net. He was actively archiving individuals' Web sites —gay and lesbian people, bi and transgender people, friends of gay and lesbian people. Tom came across the site and recruited David for PlanetOut. He was a sophomore or junior at Berkeley at that time and decided to leave—so he comes out of academia too. They launched the early service and they redid the Queer

Info Server into a product called "Net Queery," which was PlanetOut's first search engine. At the time, it was amazing that you could find gay material and gay people on the 'Net.

The other early Internet community was AOL. AOL was really the home of gays and lesbians on the 'Net. AOL was one of the first funders of PlanetOut. We were the first venture capital-funded company in the gay and lesbian market, and AOL, because they could see the traffic in the general AOL areas, said basically "Here's a great customer base that needs products. This customer needs media, community space, and goods and services that are relevant. We haven't been able to see gay and lesbian people as a group before, because people have been hiding, but here we can see them in our AOL traffic. Let's build media and products and quality stuff around that to keep those customers." Ted Leonsis [President of the Interactive Properties Group of AOL] liked that idea and invested in PlanetOut in 1996 with others. Jerry Yang [Yahoo Chief and Cofounder] from Yahoo also invested as an angel investor.

The AOL story is incredible. Steve Case [now the Chairman of AOL Time Warner] thinks that early AOL traffic was probably 15 percent gay. At one point when the 1998 Smithsonian nominations went out, PlanetOut was nominated to be in the American History Museum as a company because of the innovation we had done around community media. Steve nominated us for that, and we were accepted.

The early AOL traffic was a combination of two things. First, people could create their own gay and lesbian chat rooms and talk to each other like crazy, and many people were willing to pay large phone and ISP [Internet Service Provider] bills to do that. Secondly, there was an early pioneer, Michelle Moran. Tom and Michelle were really good friends. She was on the Digital Queers board. She was living in North Carolina at the time but within the AOL universe. With all of the gay and lesbian chat going on, she began to organize the Gay and Lesbian Community Forum. They built this service, and in 1996 changed the name to OnQ and kept growing. In the mid 1990's it was one of the best LGBTQ community hubs on the 'Net, but accessible only through AOL.

HMI: If you had to pick a few things that you'd like to do in order to make queer Internet resources better or to change the existing queer Internet resources, what would you do?

Megan Smith: One of the things we're doing is launching home pages for PlanetOut members, which is a very critical component of PlanetOut. Part of what we're doing is having different professionals talk and having varied databases of information, but it's very much about each individual's voice and what people want to say.

HMI: Launching homepages, if we look at it from an academic angle, in a sense is a way to aggregate LGBTQ academics.

Megan Smith: Sure. There could be a university-centric topic area that you could search, which would be phenomenal as a resource for people to connect with each other. Or students could talk to each other. Think of the infinite number of people—families, artists, scientists, youths, clubs, sports clubs, games clubs, whoever wants to communicate with each other—in an open-ended Web page format that ultimately will evolve the way that these things inevitably evolve.

HMI: What do you think PlanetOut may experience in terms of trends coming down the turnpike?

Megan Smith: Our member database has always been very global. By December [2000], we expect to have 900,000 registered members and several million people hitting the sites each month. We're already in over 100 countries, and as Internet connectivity increases in other places, we'll see even more people. We have discussions in different languages going on in chat rooms and message boards, and as more people come on, we'll have more voices and we'll be able to publish more material directly in those languages.

HMI: Is the diversification of languages an overall trend?

Megan Smith: It's a huge Internet trend, and it's something that we've always been a part of. We'll continue to grow as the online world grows. The most fundamental thing that PlanetOut does as a company is reduce isolation. Layered on top of that, we inform, we entertain, we connect people, and we connect resources. From an academic perspective, it's important that we not only reduce isolation within that community so that people can talk to each other more, but also that we provide a venue in which that group can do great research and create a digital research community.

HMI: Thanks Megan. Do you have any other comments about the Internet?

Megan Smith: I do think that over time, for our community, for the LGBTQ community—while it is true that there is so much hype about the Internet—I think the influence the Internet will have on our world is actually underestimated. And I think the academics are one key component of that world.

INTERVIEW: ROGER KLORESE AND WILL DOHERTY

Roger B. A. Klorese started communicating online using XTALK conferencing on Dartmouth College's time-sharing computer system in 1974, and hasn't shut up since. He is the founder and project manager of QueerNet, a nonprofit online service for GLBTQ and HIV/AIDS communities. He has worked in the computer industry since 1978, and his current "day job" is Director of Product Marketing for VMware, Inc.

Will Doherty is the founder and initial Executive Director of the Online Policy Group. He has a strong commitment to protecting rights of access, privacy, and safety on the Internet. Will also works as the online activist of the Electronic Frontier Foundation. Prior to founding the Online Policy Group, he served as the Director of Online Community Development at the Gay and Lesbian Alliance Against Defamation (GLAAD), where he focused on the online rights of lesbian, gay, bisexual, and transgender communities. He managed GLAAD's Digital Media Resource Center in San Francisco, cultivating strategic partnerships in Silicon Valley and beyond.

HMI*: Roger, could you tell us a bit about the beginning of QueerNet?

Roger Klorese: QueerNet was established in 1991. Basically, it was the result of a convergence of a number of events. I was burned out during the last days of Queer Nation (QN) and was looking for an avenue for doing queer activism, one that would provide relief from the endless in-your-face processing that QN had descended into. Most of the mailing list resources on the 'Net at the time were hosted at universities. They were constantly moving around whenever the owner of the list would move, or would need to find a new list owner when the previous owner would graduate or leave campus. One list was QN list—the mailing list for Queer Nation groups around the world and for anyone interested in participating. In one year, the list changed owners and universities three times. I wanted to create something to help us get past all the moving around of the mailing lists and how to find them. For example, GayNet was started at MIT in the mid-1980s, primarily for on-campus issues. By 1992, there was no one at MIT who would maintain it. There was an attempt to set it up at Kent State in Ohio, but there were multiple technology failures. At the time, I was working for an employer that had computer equipment available for about 30 percent of the usual cost, so I had fairly low-cost access to the necessary equipment. QueerNet was a way to create a permanent site for the various mailing lists. QueerNet currently has about 600 mailing lists and 65,000 subscribers.

Will Doherty: GayNet started out in the ARPAnet days. In 1983, I was general coordinator of Gays at MIT (GAMIT) and was working at Bolt, Beranek and Newman, an engineering lab in Cambridge, Massachusetts, that was very involved in setting up the infrastructure of the Internet. The Internet infrastructure would crash frequently, and I would have to reboot the computers that were critical to the Internet—computers that still required paper tape and had clunky readers. We did all sorts of leftist activism using the precursor to the Internet, which was rather ironic since much of the funding came from the Department of Defense.

*Interview conducted on February 12, 2001, by Alan Ellis and Kevin Schaub for HMI.

HMI: It's intriguing that you had somewhat traditional institutions such as universities and leftist groups like (QN) coming together. How well did they blend together?

Roger Klorese: Actually, a lot of the greatest passion for QN was at universities. Campus groups were really active. However, the more sex-oriented lists needed a place like QueerNet because they could not set up at universities.

HMI: Where is QueerNet physically located?

Roger Klorese: For the first seven years, QueerNet was physically located in a corner of my home office. The production servers are now at a business in a South of Market location in San Francisco.

HMI: There are probably some list owners that do not wish to use the term "queer." How does QueerNet accommodate them?

Roger Klorese: In addition to QueerNet, PLUSnet was set up in 1997 for list owners who wanted to set up lists that didn't have the "Q" word in the name. For example, a list for lesbians in the military, or sites that wanted to reach a broader audience, such as a list on infertility or for lesbian cancer survivors. A lot of the early employee groups like DECPLUS [GLBT employees at Digital Equipment Corporaton] set up lists on PLUSnet. I chose the name PLUSnet for "People Like Us," as some of the early gay employee organizations, such as DECPLUS, did.

HMI: What can you tell us about the Online Policy Group and the recent merger with QueerNet?

Will Doherty: The Online Policy Group merged with QueerNet on January 1, 2001. Our motto is "One Internet with equal access for all." The Alan Turing Program is the queer part of the Online Policy Group, and includes QueerNet and other projects, such as the Online Service Provider Assessment project, the LGBT SWAT team, and the LGBT Schools and Libraries project [for more information about these resources, see <http://www.onlinepolicy.org/program/>].

The Online Policy Group's four areas of focus are access, privacy, digital defamation, and the digital divide. Access, privacy, and the digital divide are also of interest to free speech groups. The digital defamation issue is somewhat problematic to free speech groups. We support free speech, but also believe in confronting defamation. We wouldn't advocate that an Internet Service Provider cancel the service of a hate monger, but we do want to have a discussion. So, we might ask the site owner whether or not they have thought about how this might affect a gay teenager. Sometimes we can reach into that person's conscience and he or she will remove the defamatory content. If they remove it quickly, we don't pursue the matter, but if the person or organization refuses to do so, we say "look, this is disturbing and we're going to get the word out about it." Especially if it's a corporate site, we'll ask corporate leadership how they think the LGBT community is going to respond to the negative image they are creating.

HMI: One of the great things about the Web is that it allows access for queer teenagers anywhere they might be if they are connected to the Internet.

Will Doherty: My motivation to do this work comes when I think about the transgender, queer, or questioning kid out there who is affected by what they find on the Web, both positively and negatively. But it also extends beyond teenagers; it also includes elderly LGBT people, or people with AIDS, or anyone whose connection to the outside world may be limited in some way or another. Produced by GLAAD, Loren Javier's books *Access Denied,* versions 1 and 2 provide examples of gay teenagers who say that information they found on the Web kept them from committing suicide. The Online Policy Group also has a schools and libraries project that monitors how the Internet is used in schools and libraries and what type of access LGBT students have.

Roger Klorese: In addition to the mailing lists and Web sites, the merged organization is close to becoming a domain registrar, where groups can register their domain names too.

Will Doherty: With all the mergers of Internet access and content providers, free sites are disappearing all the time, and the influence of advertisers on content that can be displayed is increasing. We will provide Web sites where people can express themselves as they wish. Currently, many of the large portals and other sites that offer free services and mailing lists can shut you down without notice if they believe your content violates their policies. The question is, do you really want to trust your data with a group that doesn't have your interests at heart? We want to provide a service by and for the communities we serve that has their interests in mind.

HMI: How could an academic researcher make use of your services?

Roger Klorese: There are a number of communities that are likely to be of interest to a researcher. For example, we have a number of lists like the lesbian studies list. Someone interested in, say, choral music as a research topic might find others with similar interests by contacting members of the mailing list for GALA choruses. Also, if you are an academic and can't start a mailing list at your institution, you can set up a list at QueerNet.

HMI: What is the process for setting up a mailing list or Web site through QueerNet?

Roger Klorese: Go to www.queernet.org, and fill out the form.

HMI: Is there a review process?

Roger Klorese: The standards are simple; the two criteria are to provide access to underserved groups and those whose content might be censored by other providers.

Will Doherty: One issue is what would we do if a group called Queer Nazis wanted a list.

Roger Klorese: I guess I would be more concerned about a group called Nazi Fag Bashers. We don't have to support groups that further bashing or violence. That would run counter to our work against digital defamation.

Will Doherty: Returning to the question about academic research, the Online Policy Group would be an excellent site for anyone studying Internet access, privacy issues, digital defamation, and the digital divide as they relate to LBGTQ communities.

NOTE

1. Gay.com and PlanetOut Plan Merger. *Gay.com*. November 16, 2000/November 28, 2000. <http://www.gaywire.net/newswire/show.cgi?20001116-170010>.

Chapter 4

Queer Studies

Sanda Steinbauer

In the past twenty years, pioneering scholars, both inside and outside universities, have forged a crucial new field of inquiry—lesbian, gay, bisexual, and transgender studies (LGBT), often referred to as queer studies or gay and lesbian studies. Their work has forced disciplines ranging from anthropology and literature to biology and art history to reassess their theoretical and political grounding, and to consider sexuality and sexual diversity as critical factors influencing social behaviors and structures. Queer studies is not limited to the study of lesbians, bisexuals, gay men, and transgender subjects. Queer studies, in short, cannot be defined exclusively by its subject, its practitioners, its methods, or its themes. Thus, queer studies is not limited to the study of queer lives and contributions: it includes any research that treats sex and sexuality as a central category of analysis. Queer studies intends to establish the analytical centrality of sex and sexuality within many different fields of inquiry, to express and advance the interests of some LGBTQ subjects, and to contribute culturally and intellectually to some contemporary LGBTQ movements. It is informed by the social struggle for the sexual liberation of LGBTQ people and by the resistance to homophobia and heterosexism.

That being said, queer scholarship suffers from a certain stasis that queer theory is determined to trouble: a symmetry that derives from prioritizing the local, the masculine, and the white subject. Taking lesbian, gay, and sexuality studies as its points of departure—and not as ends in themselves—queer theory attempts to promote scholarship about the lived experiences of sexual minorities. For many queer theorists, the definition of sexual minorities extends beyond the lesbian and gay subject, and includes BDSM practitioners, drag kings, Brazilian travestis, sex workers, intersexuals, bisexuals, transgender people, etc.

By positing *queer* as a point of departure for a critique that when ideally defined accounts for the social antagonisms of nationality, race, gender, and class, as well as sexuality, a "queer" critique seeks to occasion an intervention in what is still the formative stage of queer-theoretical engagement. Queer theory attempts to bring the projects of queer, postcolonial, and critical race theories together with each other and with a feminist analysis that itself has been an essential factor in the critique of social identity. A queer critique is conceived as a means of crossing and changing theoretical boundaries, thereby harnessing the critical potential of queer theory while deploying it beyond the realms of sexuality and sexual identity. This deployment is meant not only to inform how numerous dimensions of social experience transverse one another; it also attempts to "queer" the status of sexual orientation itself as the authentic and governing class of queer practice, thus ideally allowing queer theory to reconceptualize not just the sexual, but the social in general.

Given the diversity of gay, lesbian, bisexual, transgender, and queer subjects and the diversity of queer-theoretical engagement, novice researchers may wish to orient themselves by visiting some of the following sites which attempt to provide gateways to a large number of links. Standard search engines such as Yahoo, Lycos, and Altavista may be of limited value for locating queer studies information.

QUEER STUDIES RESOURCES

Center for Lesbian and Gay Studies
http://web.gc.cuny.edu/Clags

The Center for Lesbian and Gay Studies (CLAGS) is the queer studies center at the City University of New York (CUNY), which connects and represents the queer academic community. The Web site offers a calendar of events (usually New York City-centered); publishes announcements related to the LGBTQ communities; posts fellowships, awards, publications, syllabi, programs, and proposals; and hosts an archive of links, advocacy, and sponsor lists.

The Gay and Lesbian Review Worldwide
http://glreview.com

The Gay and Lesbian Review Worldwide (formerly the Harvard Gay and Lesbian Review) is a bimonthly journal of GLBT arts, culture, and politics. The mission of the review is "to provide a forum for enlightened discussion of issues and ideas of importance to lesbians and gay men."

Institute for Gay and Lesbian Strategic Studies
http://www.iglss.org

The Institute for Gay and Lesbian Strategic Studies (IGLSS) describes itself as "the source for timely and relevant scholarship. An independent think tank answering questions that affect the lesbian, gay, bisexual, and transgender communities, IGLSS confronts tough issues—using credible methodology to assure reliable answers. With a mix of scholarly study and rapid-response data on pressing topics, IGLSS fulfills some of the most vital research needs of the lesbian, gay, bisexual, and transgender communities, and provides leadership within the movement through informed critical analysis."[1] The Web site provides research and analysis on questions important to LGBTQ populations. The site provides access to IGLSS's journal, *Angles,* posts employment discrimination information related to queers, provides abstracts on materials written about queers, and lists fact sheets of results from original research and special reports.

The James C. Hormel Gay and Lesbian Center
http://206.14.7.53/glcenter/center.htm

The James C. Hormel Gay and Lesbian Center at the San Francisco Public Library is "a research center devoted to the documentation of lesbian and gay history and culture by collecting, preserving and providing material on all aspects of the lesbian, gay, bisexual, and transgendered experience."

The site includes an annotated bibliography of homosexuality in young adult fiction and nonfiction, resources for LGBT teens, a listing of essays on LGBT San Francisco, and an extensive collection of links to LGBTQ sites.

The Lavender Web: LGBTQ Resources on the Internet
http://www.lgbtcampus.org/resources/internet-chapter.html

David C. Barnett and Ronni L. Sanlo have complied a list of Web sites, LGBTQ newsgroups, and mailing lists to which one may subscribe via e-mail to begin connecting with information and services relating to one's campus. Their site also gives useful instructions on how to subscribe to electronic mail lists. Furthermore, they provide information on Usenet, such as what it is, why it is useful, and what subject areas it covers. (For a description of mailing lists and Usenet newsgroups in this guide, see Chapter 2, "Conducting Research on the Internet.")

Lesbian, Gay, Bisexual, and Transgender Studies and Queer Studies Programs in Canada and the United States
http://www.duke.edu/web/jyounger/lgbprogs.html

Frequently updated, this site provides a listing of courses and programs offered throughout North America in LGBT and queer studies. In addition, it offers a helpful question and answer section about such programs. You may wish to look over the listing of courses to identify a research topic or to identify researchers and academics who publish on a given topic. You can also use it to help decide on a university or college for undergraduate or graduate research in queer studies. Links and/or e-mail addresses to many of the programs listed are provided. Overall, this site is a gold mine for those interested in the current state of queer studies in the United States and Canada.

New York Public Library:Humanities and Social Sciences Library
http://www.nypl.org/research/chss/grd/resguides/gay.html

The New York Public Library's online research guide to gay and lesbian studies is an excellent resource for anyone conducting research on queer topics. The online guide provides information on general works, use of the catalogs to identify resources, and access to specific resources in the areas of humanities, history, literature, social sciences, film studies, religion, and philosophy. The site also includes a brief history of LGBTQ issues and a good bibliography of printed research guides on queer topics.

QueerTheory
http://www.queertheory.com

QueerTheory.com provides "the best online resources integrated with the best visual and textual resources in queer culture, queer theory, queer studies, gender studies, and related fields." The site is organized as a directory, with categories such as academics, bodies, identities, politics, theories, arts, and histories. Information is indexed in multiple ways: by subject, site name, author, and school. For each topic, the site lists information such as books, organizations, notable persons, and major issues. Some topics are well developed, while others include few resources (users are encouraged to add links). The site is associated with Erratic Impact's Philosophy Research Base. It provides an interesting example of how academic and commercial elements (primarily book sales) can be combined on the Web.

The Research Guide for Gay and Lesbian Studies
http://www.library.yale.edu/rsc/gayles/gayindx.htm

The Research Guide for Gay and Lesbian Studies at Yale University was created by Alan Solomon, Head Research Collection Service Librarian of the University Library at Yale. The Guide is currently maintained by Marianna McKim, the librarian for gender studies at Yale. The Web site provides a wealth of resources, links, and bibliographies.

Social Sciences and Humanities Library
http://sshl.ucsd.edu/womens_studies/gaylesb.html

The Social Sciences and Humanities Library of the University of California at San Diego includes basic references for researchers in queer studies. The site provides access to selected periodicals, paper indexes, abstracts, bibliographies of older materials, and selected nonfiction. Finally, there is a section on gays and lesbians in film and literature.

Yale University Online Research Guide
http://www.library.yale.edu/rsc/gayles/gaylib.htm

The Yale University Online Research Guide provides links to a number of sties that are useful for conducting research on queer topics in humanities.

NON-WEB RESOURCES

Center for Lesbian and Gay Studies Gender and Sexuality Studies List
http://web.gsuc.cuny.edu/clags/gendersexstudies-l.htm

This mailing list is an excellent resource and a must for any new scholar of queer theory. It keeps you updated on issues related to the study of queer studies, gender, and sexuality in academia. Refer to the URL for more information and instructions on how to join the list.

Louie Crew's Directory
http://newark.rutgers.edu/~lcrew/lbg_edir.html

Louie Crew's directory and mailing list is for LGBTQ academic researchers. The directory includes names, educational affiliations, and areas of research for over 700 university faculty and other researchers.

QSTUDY-L
qstudy@listserv.acsu.buffalo.edu

QSTUDY-L is a forum for academic discussions pertaining to queer theory. Posting about relevant conferences or publications, calls for papers, job opportunities, etc., is encouraged.

NOTE

1. *Institute for Gay and Lesbian Strategic Studies (IGLSS).* No creation date listed. IGLSS. January 19, 2001. <http://www.iglss.org>

Chapter 5

Bisexual Studies

Liz Highleyman

Many Web sites geared toward the broader LGBTQ communities include links of interest to bisexuals. But there are also sites geared specifically toward bisexuals and their allies. The first bisexual sites on the Web were personal home pages developed by members of the bi community, and many such sites are still in existence. Today, however, there are also several larger, more comprehensive bisexual sites.

The sites listed here represent some of the most prominent bi sites; most are resource sites that feature links to many other smaller, more focused sites and to the pages of bisexual groups and organizations. Several bisexual sites also offer content and links of interest to polyamorous people (for more polyamory resources, see Chapter 7, "Human Sexuality Studies").

BISEXUAL STUDIES RESOURCES

Anything That Moves Magazine
http://www.anythingthatmoves.com

This site is the Web home of *Anything That Moves,* the U.S. national magazine for bisexuals and their allies. It features a selection of articles, fiction, news, reviews, and columns from recent issues. Earlier material is being added in reverse chronological order, back to the first issues from 1991. The site also includes an events listing and links to other Web sites of interest to bisexuals.

Bi the Way
http://bitheway.org

Bi the Way, run by an anonymous bisexual activist, is among the most extensive indexes of bisexual and bi-interest Web sites. This site is rather flashy and works best with a recent version of Netscape or Internet Explorer. The site includes both bi-specific and general LGBTQ links. The content of the linked sites varies widely. Some links go to articles (for example, an essay on bisexuality and psychology, reprints of Bisexual Resource Center pamphlets), others go to larger bi Web sites, and a few go to sexually explicit pages. The site includes links to the most prominent U.S. and international bisexual organizations, e-mail lists, and chats, but also links to many eclectic bisexual sites that are not included in the other major bi resource listings. The site serves as a jumping-off point for many bi-oriented Web rings. One nice feature of this site is a search section that brings together in one place an extensive collection of LGBT and women-oriented search engines.

BiCafe
http://www.bicafe.com

BiCafe.com is widely regarded as the most highly trafficked bi-oriented Web site, with a focus on building an online community for "bisexual, bi-friendly, and bi-curious" people. Although a paid subscription is required for full access, much information is available in the site's free zone, including editorials, news, columns, informational essays, and a calendar of events (the "Bi Agenda"). The paid area includes personals, e-mail, and forums. The site also has a PolyBi section (free) and an online store selling bi-oriented merchandise. Much of the content on BiCafe.com is available in both English and French.

Bi.org
http://www.bi.org

Bi.org is a bisexual resource site with an international focus. It includes links to a large number of bisexual groups and organizations by country, Usenet newsgroups, electronic mailing lists, and IRC channels from around the world (U.S. sites are mostly handled by Bisexual.org, described as follows). The site is the home of the UK's *Bi Community News* magazine. Bi.org contains listings of bi and bi-friendly links in many topic areas—for example, academic resources, health and disabilities, and erotica—arranged in a familiar directory structure like that used by Yahoo and the Open Directory Project. In addition to bi-specific sites, Bi.org includes gay, lesbian, bisexual, polyamory, transgender, and women's links. It features linked lists of books, music, films and magazines of interest to bisexuals, as well as calendar and conference listings and international bisexual news. It also provides a search engine and free e-mail accounts.

Bisexual.org/Bisexual Options
http://www.bisexual.org

The Bisexual.org Web site is a good starting point for finding bisexual resources. It includes links to bi-oriented Web pages, a city-by-city listing of U.S. bi groups and organizations by region, and links to bi-focused Usenet newsgroups, electronic mailing lists, and IRC channels. In a shared effort, Bisexual.org deals with U.S. resources, while Bi.org (described previously) handles similar listings for groups and resources outside the states. Bisexual.org also includes links to commercial sites, a listing of bisexual conferences, a comprehensive list of bi/polyamory/sex-positive books (with reviews), films, and links to bisexual magazines. Bisexual.org was formerly known as Bisexual Options, and includes information about Fritz Klein's book of the same name and his Klein Sexual Orientation Grid. The site is hosted by and includes information about the American Institute of Bisexuality, which provides grants to fund bisexual projects.

The Bisexual Resource Center
http://www.biresource.org

Boston's Bisexual Resource Center (BRC)—which has a long-standing role as a clearinghouse for bisexual information and publishes the comprehensive biannual *Bisexual Resource Guide*—also maintains a large collection of Web resources for bisexuals and their allies. The site includes links to articles, conferences, radio and TV shows, bisexual candidates for electoral office, health information, opportunities for writers, activist resources, newsletters, and more. The site features the full text of the BRC's series of informational pamphlets on topics such as biphobia, how to start a bi support group, the history of the bisexual movement, conference budgeting, and responsible nonmonogamy. The BRC site is the gateway to several Boston bisexual organizations, including the Boston Bisexual Women's Network. The site also has an online store selling books, music, and films of interest to bisexuals.

Bisexual Youth
http://www.biyouth.org

This site, a project of the Bisexual Resource Center, features links to a variety of resources for queer youth. Most of the resources for parents, teachers, and students have a gay or LGBT—rather than a specifically bisexual—focus. The bi youth resources section includes links to local groups for young bisexuals, online forums, and an entry point to the Bi Youth Web Ring. The site also features an article on "Crisis Counseling and Support for Bisexuals" from the Gainesville Bisexual Alliance, and "I Think I Might Be Bi (Now What Do I Do?)," a pamphlet for young people questioning their sexuality.

NON-WEB RESOURCES

Brown University Bi Lists
listserv@browvm.brown,edu

Although there are now many electronic mailing lists and discussion groups for bisexuals, the Brown University lists were among the first and remain among the best known and most widely used. These lists include **bisexu-l** (a free-form, high-traffic list for bisexuals and those interested in bisexuality), **bifem-l** (a discussion group for bi women), **biact-l** (a list for bisexual activists), and **bithry-l** (a lower-traffic discussion group loosely focused on the theory—or perhaps more accurately, the philosophy—of bisexuality). To subscribe, send e-mail to <listserv@browvm.brown.edu>, and request a subscription to the list(s) of your choice.

BiFem Net
http://www.bifem.net

BiFem Net is another good resource, especially for bi women. It serves as the gateway to the BiFemLounge and subscriptions to several U.S. and international electronic mailing lists.

Soc.bi

The primary Usenet newsgroup for bisexuals is **soc.bi.** This group receives a large amount of traffic covering a wide range of subjects of interest to bisexuals, the bi-curious, and their friends and allies. Like many newsgroups, **soc.bi** is unmoderated and prone to periodic "flame wars." Several other newsgroups also contain bi-related content, including several adult-oriented forums in the **alt** hierarchy.

Chapter 6

Transgender and Intersex Studies

Alan Ellis
Liz Highleyman

Increasingly, academic researchers are turning their attention to issues of concern to the transgender community. The establishment of the *International Journal of Transgenderism* in 1997 is one indication of the growing academic interest.

TRANSGENDER AND INTERSEX STUDIES RESOURCES

Above and Beyond
http://www.abgender.com

Above and Beyond is a searchable database of information and resources for the transgender community. The site is a directory with over 2,400 pieces of content arranged in categories such as fiction, medical, news, and personal Web pages. There are links to chats and over twenty newsletters from clubs and organizations around the world. The article section contains nearly 100 articles dealing with gender theory, legal issues, and other topics of interest to researchers. Above and Beyond is also the jumping-off point for several different gender Web rings.

FTM Information Network
http://www.ftminfo.net

The FTM Information Network Web site provides a range of information for female-to-male transgendered people, including news, legal issues, essays, medical information, personal stories, and links to other sites of interest.

FTM International
http://www.ftm-intl.org

FTM International provides one of the best Web sites for female-to-male transgendered people, with lots of good information and resources. The site offers news, historical information, FTM biographies, gender-related legal information, and transition information (including standards of care, hormones and surgery, and social aspects of transitioning). It also features access to an archive of FTM International newsletters, several bibliographies, a list of relevant books and publications, information about meetings and events, and links to other Web sites of interest.

Gender Education and Advocacy
http://www.gender.org

Gender Education and Advocacy (GEA) is a national organization formed in January 2000, growing out of the American Educational Gender Information Service (AEGIS). GEA's mission is to focus on "the needs, issues and concerns of gender variant people in human society." The GEA Web site is one of the best resources for transgender people and those interested in gender issues. It features the archives of Gender Advocacy Internet News (GAIN; see Non-Web Resources, as follows) and information and resources related to a wide range of topics such as child custody, health, TS/TG teachers, and employment issues. The site provides links to GEA projects including the National Transgender Library and Archives and the "Remembering Our Dead" memorial. One of the most useful features of the site is the Trans-Portal (http://www.gender.org/resources/links.html), an exhaustive directory of links covering everything from activism to drag to transgender youth to intersexuality. (An interview with GEA board member and Webmistress Gwendolyn Ann Smith appears at the end of this chapter.)

GenderPAC
http://www.gpac.org

GenderPAC is a national advocacy organization that works "to ensure every American's right to their gender free from stereotypes, discrimination and violence, regardless of how they look, act or dress, or how others perceive their sex or sexual orientation." The group has recently broadened its focus beyond transgender and transsexual activism to include all people who are marginalized on the basis of their gender presentation or behavior. GenderPAC's work includes congressional lobbying, education, and litigation. Their Web site includes news and information about issues on which GenderPAC is currently focusing, such as job discrimination, hate crimes, and gender law.

International Foundation for Gender Education
http://www.ifge.org

The International Foundation for Gender Education (IFGE), founded in 1987, describes itself as "a leading advocate and educational organization for promoting the self-definition and free expression of individual gender identity." IFGE is "an information provider and clearinghouse for referrals about all things which are transgressive of established social gender norms." The group provides information and services for MTF and FTM transgender and transsexual people and cross-dressers. The IFGE Web site features news, a calendar of events, lists of gender-related books and movies, tables of contents from *TG Tapestry* magazine, full-text articles from the *IFGE Newsletter*, and links to local support groups and national gender organizations.

The International Journal of Transgenderism
http://www.symposion.com/ijt

The *International Journal of Transgenderism* is a peer-reviewed academic online journal. The editorial board consists primarily of individuals holding MD and PhD degrees. The Web site provides free access to electronic books, abstracts, full-text articles, historical papers, and printed digests. Recent online articles focus on various topics such as factors that influence an individual's decisions when considering female-to-male genital reconstructive surgery, and gender role reversal among postoperative transsexuals. A recent special issue is titled *Transgender and HIV: Risks, Prevention, and Care*. The full text of all articles is available online.

Intersex Society of North America
http://www.isna.org

The Intersex Society of North America (ISNA) is "an education, advocacy, and peer support organization which works to create a world free of shame, secrecy, and unwanted surgery for intersex people." The site is the primary Internet resource for intersexed people. It covers ISNA's history, news, public policy statements, and statistics about the frequency of intersexuality.

The site defines intersex people as "individuals born with anatomy or physiology which differs from cultural ideals of male and female." There is also information in Spanish and an extensive bibliography and publications list, including some links to full-text articles. In addition, there is an archive of ISNA's newsletter, *Hermaphrodites with Attitude*, dating back to 1994.

It's Time, America
http://www.tgender.net

It's Time, America (ITA) is a grassroots civil rights group formed in 1994 "to secure and safeguard the rights of all transgendered and gender variant persons." ITA has local chapters in several U.S. cities. The ITA Web site includes information about the group and a directory of contacts, press releases about gender issues that the group works on, and a library of provocative articles including "TS Feminism and TG Politicization," "The Catholic Church on TSism," and "Silence Will Not Protect You."

Jacob Hale's Rules
http://www.sandystone.com/hale.rules.html

This site lists "suggested rules for non-transsexuals writing about transsexuals, transsexuality, transsexualism, or trans _____."

National Transgender Advocacy Coalition
http://www.ntac.org

The National Transgender Advocacy Coalition (NTAC) bylaws list a variety of purposes for the organization, including "to engage in efforts to provide education with respect to issues related to gender variance and public gender presentation, and to act as an information resource." In support of these goals, NTAC's site offers a number of essays and commentaries on transgender issues and, occasionally, requests for individuals to participate in transgender research projects.

Transgender Forum
http://www.transgender.org

The Transgender Forum offers free Web space to transgender-oriented community organizations throughout the United States and Canada. If you are researching transgender issues, you may wish to begin here to assess current issues and concerns of the transgender community. The site includes multiple links to transgender online and print resources.

Transgender Law and Policy
http://transgenderlaw.org

This recently developed site features news, articles, cases and briefs, and information about nondiscrimination laws, employer and college policies, and hate crime laws as they relate to transgender, transsexual, and gender-variant persons. It also includes a collection of links to other resources.

NON-WEB RESOURCES

Gender Advocacy Internet News (GAIN)
gain@gender.org

GAIN is a free transgender/transsexual (TG/TS) Internet news service edited by Penni Ashe Matz, a transperson living in the greater Boston area. The news service is sent as a digest twice weekly. This is an indispensable resource if you are researching issues affecting the TG/TS community. An archive of past news is available at <http://gender.org/gain>. To subscribe to or unsubscribe from GAIN, visit <http://www.tgender.net/mailman/listinfo>.

Transgen
http://songweaver.com/lists/transgen.html

The Transgen list, hosted at Brown University, was one of the earliest transgender mailing lists, started in 1990. Discussions span a wide range of topics of relevance to transgendered people and those interested in gender issues. For information about the list, visit the URL. To join the list, send an e-mail to transgen-request@listserv.brown.edu. For archives dating back to 1997, see <http://listserv.brown.edu/archives/transgen.html>.

Yahoo Groups Transgender Online Community
http://groups.yahoo.com/community/transgender

This transgender discussion group is one of the most active and is open to anyone interested in transgender issues.

INTERVIEW: GWENDOLYN ANN SMITH

Gwendolyn Ann Smith is a board member of Gender Education and Advocacy, Webmistress for the Gender.org Web site, the community host of Gay.com's Transgender Gazebo, and columnist for the *Bay Area Reporter* covering transgender issues.

HMI*: Please describe your background and how you became interested in computers and the Internet.

Gwen Smith: I grew up in a city in Southern California. Now, one might think that being in proximity to Los Angeles would mean being in a mecca of transgender resources, but I discovered that this was not the case. My local library had nothing, nor did my college library. It was only after a lot of hard work that I begun to dig up a few scant resources, and I had even fewer opportunities to meet others such as myself.

My background was in graphic design, and when I first sat down in front of a Macintosh at the tail end of the 1980s, I discovered a tool that reshaped the way I approached design. I jumped at the chance to learn more about computers, and discovered that I seemed to intuitively understand how they worked.

I had just started to get involved in the transgendered world in early 1993 when I received my first (but definitely not my last!) America Online (AOL) membership diskette in the mail. I decided it might be fun to try it out, as the idea of people getting onto their computers and

*Interview conducted via e-mail, February 2001, by Liz Highleyman for HMI.

being able to share with each other sounded interesting. It didn't take long for me to get hooked, especially when I found my first online chat room for transgendered people, called TV Chat.

By late 1993, a few friends of mine decided to band together into an informal organization called the American Online Gender Group, or AOLGG. Our goal was to get AOL to remove "transsexual" and "transvestite" from the list of "vulgar words" used by their Terms of Service department. The Terms of Service determine what is and isn't allowed on their system, and failure to follow their Terms of Service can lead to the termination of your account. We also had vague hopes of seeing the formation of an online forum for transpeople — a dream that became reality with the assistance of the Gay and Lesbian Community Forum on AOL, which provided us with space and resources to open up the Transgender Community Forum, or TCF.

Nothing lasts forever, and the TCF has since become the Transgender Gazebo on Gay.com, for which I am community host. The resources and community we once provided on AOL are now on Gay.com and Gender.org.

HMI: What are some of the Web projects with which have you been involved?

Gwen Smith: In mid-1995, a business acquaintance of mine suggested that I get involved with the emerging World Wide Web, and I started to provide transgender resources and information in that medium. This led to a number of Web projects, from trans-related business and organizational sites, to more artistic sites like "Online Alchemy: The Art of Loren Cameron" <http://www.lorencameron.com>.

Of these projects, the one I consider most important is "Remembering Our Dead" [http://www.gender.org/remember], a Web-based memorial to individuals killed due to anti-transgender violence or prejudice. It was nominated for a GLAAD Media Award and discussed in several LGBTQ publications. More importantly, the information I have provided on the site has been used in numerous transgender activist circles to facilitate the understanding that violence against transgender-perceived people is a major problem, with eighteen people reported killed during 2000.

HMI: What role does the Internet play in informing, empowering, and supporting the transgender community?

Gwen Smith: The 'Net plays a major role in the transgender community. We are one of the first communities, in my opinion, to take full advantage of the cyber world, using it as a way of educating both each other and those outside of our experience.

It wasn't until the early 1990s that transgender activism as we know it today really began to take hold, and the transgender community has the Internet to thank. Some of the protests in the early and mid-1990s, such as those concerning the Brandon Teena murder and the Sean O'Neil court case, were organized and promoted largely over the 'Net. This still happens today. In November 2000, Day of Remembrance vigils were held in fifteen different cities across the country, coordinated via the 'Net.

One of the more important roles played by the Internet is keeping people informed of what happens worldwide. In the transgender world pre-Internet (or, more precisely, before the 'Net really took off and became widely used), communication was largely based on sporadic periodicals and word of mouth. The Internet allowed instant communication.

HMI: Does the Internet have unique features that make it especially beneficial for the transgender community?

Gwen Smith: There are two features that make the Internet especially tantalizing for transgendered people, both related. For one, you can be mostly anonymous, allowing one to be themselves with a lot less risk. It also allows one to try on new identities and "shop around" for what feels right. Both of these make the 'Net the world's largest gender playground.

HMI: What are some major mailing lists, Web sites, and other resources for the transgender community? What were some of the earliest transgender Internet resources, and how have these evolved?

Gwen Smith: The 'Net for transgendered people has changed dramatically from the days of early BBS systems like Cross-Connection and such. It was these, as well as early pioneers like the TCF and

GenderLine (on CompuServe) that paved the way. These are all but gone now, but they did show us what was possible.

Another early resource was the transgen mailing list, out of Brown University [transgen@listserv.brown.edu]. It is among the few early mailing lists still going today.

Nowadays, one of the larger mailing lists out there is the GAIN [Gender Advocacy Internet News] list, which is a news and information list for the community. A scattering of other mailing lists and newsgroups are also out there, focused on FTMs, post-ops, and so on.

Things really seemed to get going with Web sites, from some of the large commercial ventures like TGForum [http://www.tgforum.com], Transgender Gazebo [http://content.gay.com/people/trans_gazebo] and ABGender [http://www.abgender.com], to informational sites like Gender.org, FTM Informational Network [http://www.ftminfo.net] and TS Roadmap [http://www.tsroadmap.com] all the way down to the thousands of personal transgender Web sites out there.

HMI: Do you have any anecdotes about how the Internet has affected the lives of transgendered people?

Gwen Smith: That covers a lot of ground. I've seen people new to the trans portions of the Internet, who are scared, lonely, and confused. Over time I've seen these same people mature and grown into very happy people. It has made me happy to see this happen, and it is that sort of experience that keeps me involved.

I recall back in 1993, sitting around in a meeting with a few of the AOLGG founders, discussing what we felt was a "blue sky" idea—the thought of actually making an area on the Web for transgendered people to communicate with each other and to share our experiences. It all seemed like an interesting fiction, but nothing we would see in at least the next ten years. Little did we know at the time what an explosion we would see in the popularity of the Internet and Web, which made our "blue sky" something incredibly common.

HMI: Is there anything else you'd like to add regarding the Internet and transgender community?

Gwen Smith: What will be the next "blue sky" for our community? That's hard to say. There are new voices coming up in the ranks who are exploring gender presentations that I never could have imagined, and who are using the 'Net to deliver their message in new and unique ways. Technologies like Flash animation and Adobe Acrobat, coupled with the growing availability of digital cameras and high-bandwidth connections, are making it increasingly easy to present materials that would not have been possible five years ago.

Chapter 7

Human Sexuality Studies

Liz Highleyman
Sanda Steinbauer

For many scholars and activists, the term "queer" reaches beyond gay, lesbian, bisexual, and transgender to encompass a range of human sexual identities, preferences, and practices located on the margins of the social and cultural mainstream. The study of human sexuality spans this full range of sexual identities and practices, as well as looking at how sexual identities and practices, reproduction, and family relationships differ among various societies and cultures. In this chapter we include resources related to human sexual variation, sex education, and different sexual minority or alternative sexuality communities. These topics have in common the fact that they cross gender and sexual orientation boundaries.

For the purposes of this chapter, "BDSM" is used as an umbrella term used to refer to sadomasochism (SM), dominance and submission (DS) and other activities that involve erotic power exchange. "Leather" refers to similar practices, but has a stronger (or at least older) association with a specific—traditionally gay male—community and lifestyle.

"Polyamory" is a relatively new term for multiple-person relationships. More than "nonmonogamy," "polyamory" suggests consensual relationships that may be (but need not be) long-term and emotional as well as sexual. Polyamory refers to both relationships that involve more than two persons simultaneously (for example, a triad) and open relationships in which partner(s) have sexual relationships with person(s) other than their primary partner.

Over the past decade "sex work" has become an umbrella term used to refer to those who work in the "sex industry," that is, in jobs that involve sexual services or entertainment. Some researchers and activists use the term only for prostitutes; others include different types of providers such as exotic dancers (strippers), dominatrices, and porn actors; while still others extend the scope to include erotica writers, manufacturers of sexual devices, support staff and others who do not directly perform sexual labor.

For many sexual minority and alternative sexuality communities and practitioners, the Internet has provided unique opportunities for connection, visibility, and education. Many people who may have once thought themselves alone in decades past can now find information on the Internet about almost any conceivable sexual practice. BDSM practitioners, polyamorists, and sex workers have all developed strong, activist communities in recent decades, aided in no small measure by the Internet, especially e-mail and the Web. The Internet also provides a great deal of materials and resources for researchers who seek to study human sexuality in all its diversity.

GENERAL SITES

Altsex.org
http://www.altsex.org

The Altsex Web site is "dedicated to the exploration of the miracle of human sexuality, in all its wonder and diversity." The site includes articles on polyamory, bondage, sadomasochism, transgender issues, and features links to other sex-positive online resources.

The Kinsey Institute
http://www.indiana.edu/~kinsey/journals.html

The Kinsey Institute's Web site provides a selected list of journals that have a strong focus on human sexuality or that have a special relevance to current research projects at the Kinsey Institute. This is an excellent source for journals in the field of human sexuality. Included are such peer-reviewed, interdisciplinary journals such as the *Electronic Journal of Human Sexuality,* published by the Institute for Advanced Study of Human Sexuality in San Francisco; and *GLQ: A Journal of Lesbian and Gay Studies,* published by Duke University Press.

Included on The Kinsey Institute Web site are compiled bibliographies as well as links to other bibliographies by researchers, health professionals, and librarians (http://www.indiana.edu/~kinsey/biblio. html). Links to selected online bibliographies include the Bibliography on Human Sexuality, located at <http://www.library. upenn.edu/ vanpelt/guides/humsex.html.>

The site also features free access to databases that relate to the study of human sexuality, gender, and reproduction (http://www.indiana.edu/~kinsey/bdatabases.html). One such database, Human Relations Area Files (HRAF) from Yale University, contains full-text ethnographies of some 400 ethnic, cultural, religious, and national groups worldwide. In addition, you will find links to the Minority Health Research Catalog. The site also provides a listing of other U.S. and international research centers and special collections related to the study of sexuality (http://www.indiana.edu/~kinsey/centers.html). In this section you will find links to research and data concerning particular sexuality-related issues, such as reproductive health; HIV/AIDS research; data on Americans' attitudes toward sexuality and their sexual behaviors for over twenty years; information on the nudist movement; and information from the British Library, which holds a large collection of erotic literature. This site also includes information on forced prostitution, child prostitution, and early or forced marriages.

National Coalition for Sexual Freedom
http://www.ncsfreedom.org

The National Coalition for Sexual Freedom (NCSF) is perhaps the foremost organization in the U.S. today fighting for the rights of sexual minorities. The group's mission is "advocacy and lobbying to promote tolerance of sexual minorities and of those who engage in, write about, or study sexual minority practices among consenting adults." Although focused on those who practice BDSM, NCSF supports other sexual minorities as well. The NCSF Web site is a good source for news and information about civil rights, and legal and political issues concerning sexual minorities. It's Document Library includes basic educational information, policy statements, and amicus briefs.

Sexuality Information and Education Council of the United States
http://www.siecus.org

Sexuality Information and Education Council of the United States (SIECUS) describes itself as "a national, non-profit organization which affirms that sexuality is a natural and healthy part of living." SIECUS "develops, collects, and disseminates information, promotes comprehensive education about sexuality, and advocates the right of individuals to make responsible sexual choices."[1]

The SIECUS Web site offers an extensive list of links to other organizations that focus on sexual issues and concerns, including activism; facts and statistics; foundations; gender; government; health and wellness; HIV/AIDS and STDs; libraries, publications, references and resources; parents and teens; population organizations; religion; reproductive health and family planning; schools with HIV/AIDS education or sexuality education classes and/or programs of study; sexual orientation; sexuality organizations; and women's health.

Society for Human Sexuality
http://www.sexuality.org

The Society for Human Sexuality provides one of the best and most extensive Internet resources on a wide range of sex and sexuality topics, and serves as a Web library of alternative sexuality information. It includes in-depth information on safer sex, STDs, sexual practices, BDSM, sex work, polyamory, and Tantric sex. It features articles, essays, and interviews with well-known personalities in the alternative sexuality communities. See <http://www.sexuality.org/ftpsite.html> for a library index.

Society for the Scientific Study of Sexuality
http://www.ssc.wisc.edu/ssss

The Society for the Scientific Study of Sexuality (SSSS) describes itself as "an international organization dedicated to the advancement of knowledge about sexuality. It is the oldest organization of professionals interested in the study of sexuality in the United States. SSSS brings together an interdisciplinary group of professionals who believe in the importance of both the production of quality research and the clinical, educational, and social applications of research related to all aspects of sexuality."[2]

The site includes links to the publications of the society which include *Sexual Science* (formerly *The Society Newsletter*), the *Annual Review of Sex Research; Journal of Sex Research; Educational Opportunities in Human Sexuality;* and a brochure titled *What Sexual Scientists Know.*

BDSM/LEATHER/FETISH RESOURCES

Different Loving
http://gloria-brame.com/diflove.htm

The Different Loving Web site, the online companion to the seminal book of the same name, features several full interviews from the book with well-known figures in the BDSM/leather/fetish communities, including psychologist Dr. William Henkin, professional dominant Cleo Dubois, body modification master Fakir Musafar, and gay male BDSM pioneer Joseph Bean. The site also includes an extensive collection of over 1600 links to BDSM resources ranging from clubs and organizations to safer sex and political information, plus links for those interested in specific fetishes or practices.

The Eulenspiegel Society
http://www.tes.org

The Eulenspiegel Society (TES), based in New York City, is the oldest continuous organization in the U.S that supports the practice of BDSM. The Web site offers a valuable resource for those doing research on BDSM practices, community, and activism. It has many links to resources, safety issues, and health care related to the practice of BDSM. There are also links to other leather community organizations, the Leather Archives and Museum, and leather events.

Leather Archives and Museum
http://www.leatherarchives.org

The Leather Archives and Museum (LA & M), "based in Chicago and serving the world," is the foremost repository of BDSM/leather/fetish history. The LA & M collects materials relevant to BDSM practitioners of all genders and orientations, but most of its current collection relates to the gay male leather community. Although its Web site is currently not extensively developed, the physical archives are a major resource for those researching the history of BDSM/leather/fetish communities. Work is ongoing to make selected exhibits, a timeline, and more information available via the Web.

The Leather Page
http://www.leatherpage.com/

The Leather Page, compiled by International Mr. Leather 1996 Joe Gallagher, is a good source of news about issues affecting the BDSM/leather/fetish communities, as well as news relating to LGBT communities. It includes links to columns by well-known community activists such as Mister Marcus Hernandez, Lolita Wolf, and Robert Davolt. The site also features essays and speeches on topics ranging from the history and evolution of the leather scene, to legal and political issues affecting BDSM practitioners.

LeatherQuest
http://www.leatherquest.com

LeatherQuest, a project of Chicago-based Club Quest and the Suttle Leather Cares Foundation, is a good source of a range of information and resources for and about the BDSM/leather/fetish communities. It includes a news section; a legal resource center featuring an archive of court cases related to BDSM, censorship, sodomy, and related sexuality issues; nearly 100 educational articles on everything from political to legal to medical information, including several basic introductory articles on BDSM practices, community issues, and safety. The Gathering Places and Vendors section lists BDSM organizations and events in all fifty states and a growing number of countries.

LeatherWeb
http://www.leatherweb.com/

LeatherWeb, a "global leather village" on the Internet, provides a wealth of information for those interested in BDSM/leather/fetish lifestyles. It features news, educational material (including definitions and safer sex information), and a listing of resources and advocacy groups. It is divided into three sections focusing on gay/bi men, lesbian/bi women, and heterosexual/bi BDSM practitioners.

Society of Janus
http://www.soj.org

The Society of Janus, founded in San Francisco in 1974, is a pansexual BDSM organization for people of all genders and sexual orientations. The group's Web site includes membership and event information, and also has content of interest to non-local readers, such as an extensive listing of BDSM resources (including definitions, FAQs, articles, pamphlets, and manuals), a list of BDSM books, and links to other organizations.

POLYAMORY RESOURCES

The Poly Page
http://www.altsex.org/poly.html

The Poly Page presents a nice collection of polyamory resources. Features include a short glossary of relevant terminology, links to articles and other polyamory Web sites, and the PolyBureau (http://www.altsex.org/polycol.html), a collection of answers to frequently asked questions about polyamory and practical advice for polyamorists written by Alan Wexelblat.

Polyamory.com
http://www.polyamory.com

Polyamory.com describes itself as a site providing "resources for Polys and those who love them." It includes news as well as links to organizations, books, and other resources. It is also an entry point to the polyamory web ring.

See <http://www.polyamory.org/Howard/> for a collection of polyamory links, including organizations, books, magazine and newspaper articles, songs and movies, religious and legal links, and links related to sociology, psychology, and biology as they relate to polyamory.

Polyamory Frequently Asked Questions
http://www.cs.ruu.nl/wais/html/na-dir/polyamory/faq.html

On this Web page, you will find a list of frequently asked questions and answers about polyamory and multiple-person relationships.

SEXUAL HEALTH AND EDUCATION SITES

There are a number of good, queer-friendly—though not queer-specific—sexual health and education sites on the Web. The sites that follow provide information about issues such as safer sex, sexual orientation, contraception, HIV/AIDS and other sexually transmitted diseases, and drug and alcohol education.

Coalition for Positive Sexuality
http://www.positive.org

The Coalition for Positive Sexuality (CPS) is a nonprofit, grassroots activist organization that provides nonjudgmental information on sex and sexuality for teens and young adults. The CPS site features articles about how to tell if you're gay/lesbian/bisexual, and good sexual health and safer sex information for people of all orientations, in both English and Spanish.

Go Ask Alice
http://www.goaskalice.columbia.edu

This site, from Columbia University's Health Education Program, features information about sexuality, sexual health, relationships, emotional health, and substance use. The site is aimed at college students, but its friendly, nonjudgmental information is appropriate for young adults in general.

Safer Sex Pages
http://www.safersex.org

The Safer Sex Pages present everything you ever wanted to know about safer sex, including information about HIV and oral sex, and information for women who have sex with women. The site also features a selection of research reports on condoms and the effectiveness of safer sex education.

San Francisco Sex Information
http://www.sfsi.org

San Francisco Sex Information (SFSI) is a unique free switchboard that has been providing information about sex and sexuality for twenty-five years. Although begun as a phone hotline, SFSI now provides information and referrals by e-mail as well. SFSI's Web site provides comprehensive, sex-positive information for people of all genders and orientations.

SEX WORK RESOURCES

Commercial Sex Information Service
http://www.walnet.org/csis

The Vancouver-based Commercial Sex Information Service (CSIS), a site maintained by volunteers, is a clearinghouse of information about laws, sexual health, commerce, and culture as they relate to sex work. The culture section includes a collection of works of art featuring prostitutes and a bibliography of books, films, and plays about sex workers. The health section currently includes safer sex tips and research abstracts about sex workers, sex work, and sexual health; research on prostitution and violence is slated to be added in the future. The legal section contains government reports, articles analyzing the legal system's impact on sex workers, transcripts of trials involving sex work-related laws, and advocacy listings. Much of the information has a Canadian focus, but many of the issues covered are of broader interest.

HOOK Online
http://www.hookonline.org

HOOK Online is a not-for-profit project by, for, and about men working in the sex industry. It features original articles, interviews with male sex workers, an arts section, and health and finance forums. A resources section provides information about health centers, money management, and sex worker organizations. There is also a well-developed set of links to other Web resources. The site is probably the best existing resource specifically for and about men in the sex industry.

PENet: Prostitutes' Education Network
http://www.bayswan.org/penet.html

PENet, sponsored by the Bay Area Sex Worker Advocacy Network (BAYSWAN), is an "information service about legislative and cultural issues as they affect prostitutes and other sex workers." The site features information and resources for sex workers, activists, educators, and students on issues such as decriminalization, human rights, violence, pornography, art, health, and current trends in legislation and social policy in the United States and internationally. PENet's site includes links to several sex workers' rights organizations. It also includes a section on resources for students studying sex work and a listing of videos for educational purposes.

Sex Workers Alliance of Vancouver
http://www.walnet.org/swav

The Sex Workers Alliance of Vancouver (SWAV) presents a nice site with information and resources for female and male prostitutes and sex workers' rights advocates. Features include news, legal information, myths about prostitution, articles, and health information cards on topics such as hepatitis and the risks of nonoxynol-9.

World Sex Guide: A Research Project About Prostitution Worldwide
http://www.worldsexguide.org

The World Sex Guide is a collection of prostitution-related information covering most countries of the world. Many articles were originally posted to the Usenet newsgroup **alt.sex.prostitution** and have subsequently been archived here. Others were sent by e-mail. The site is useful for research on sex workers and the sex industry. The site includes international resources and a glossary of terms related to the professions within the sex industry. Anonymous volunteer researchers contributed all of this material. The site does not accept advertising money.

NON-WEB RESOURCES

Alt.sex.bondage and Related Resources

Alt.sex.bondage, more commonly known as **asb,** was the first major Usenet newsgroup to deal with BDSM/leather/fetish topics. Over the years, **asb** became overwhelmed with "flame wars," spam, and sheer volume. Those who desired a more serious Usenet discussion of BDSM created **soc.subculture.bondage-bdsm** (or **ssbb**). You can read the ssbb FAQ—an extensive introduction to BDSM terminology, practices, myths, and more—at <http://www.unrealities.com/adult/ssbb/faq.htm>. Feeling that their interests were not adequately represented on the main pansexual newsgroup, gay and lesbian members formed a mailing list called gl-asb, which is now hosted by QueerNet.

Alt.polyamory

Alt.polyamory is a Usenet newsgroup for people interested in talking about polyamory and related topics.

The Poly List
http://www.polyamory.org/Howard/poly.html

The Poly List, formerly known as the "triples" list, is one of the longest-running Internet discussion forums for people interested in multiple-partner relationships (the name was changed because it is not, and never was, just about threesomes).

Sex Work Mailing List
sex-work@hivnet.ch

The sex work mailing list was set up to facilitate discussion of sex work issues as they relate to HIV/AIDS. Today, the list is managed by Health and Development Networks and includes a broader range of topics including legalization, global trafficking, and medical information. The majority of list members are advocates, social workers, medical professionals, academics, and the like who work with or do research that involves sex workers (primarily prostitutes). You can join the list and read current and archived messages at <http://www.hivnet.ch:8000/topics/sex-work/>.

NOTES

1. Sexuality Information and Education Council of the United States. (No date listed). *Organizational description*. Los Angeles: Author. Retrieved January 19, 2001 from the World Wide Web: <http://www.siecus.org>.

2. Society for the Scientific Study of Sexuality. (no date listed). *Organizational description*. Retrieved January 19, 2001, from the World Wide Web: <http://www.ssc.wisc.edu/ssss>.

Chapter 8

Liberal Arts and the Humanities

Alan Ellis
Mark Menke

The majority of queer studies researchers and theorists currently are associated with the liberal arts and the humanities. As a result, the online resources listed in Chapter 4, "Queer Studies" are relevant to those conducting queer research in these disciplines. This chapter adds to those resources by focusing on additional general resources, as well as specific resources in history, ethnic studies, and religious studies.

HISTORY

Canadian Lesbian and Gay Archives
http://www.web.net/archives

For those wishing to research the history of the LGBTQ communities in Canada, this is an excellent site. The site's mandate is to collect and maintain information related to gay and lesbian life in Canada, but the site also has information from other countries as well. The Canadian Lesbian and Gay Archives gathers information to secure it for the future and to make that information available to the public for education and research. In addition to its online resources, the archives staff offers assistance to individuals such as students, artists, journalists, lawyers, and filmmakers.

Gay, Lesbian, Bisexual, and Transgender Historical Society of Northern California
http://www.glbthistory.org

The GLBT Historical Society of Northern California offers a multitude of online resources for those conducting research in GLBT history, including information on how to conduct oral histories and information about specific oral history projects with which the society is involved. Sample oral history projects include, *A Credit to Her Country* (about lesbians in the military), *Elder Lesbians* (histories of lesbians over the age of sixty), *McCarthy Era* (interviews of gay men and lesbians who suffered as a result of the McCarthy hearings of the 1950s), and *North Beach* (a history of gay culture in the North Beach district of San Francisco in the 1950s and early 1960s).

Gayhistory.com
http://www.gayhistory.com/

Gayhistory.com describes itself as "an introduction to the stories and people of modern gay history (1700-1973)." The site includes articles on a variety of historical topics, including a graphical timeline of important events from 1700 to 1900, an extensive bibliography arranged by author and subject, and a glossary of words unique to modern gay history. The site also provides an extensive list of links to other LGBTQ history Web sites. Another useful timetable of events of note for LGBTQ studies that begins in the year 1290 can be found at <http://www.sbu.ac.uk/~stafflag/timetable.html>.

Gay History and Literature
http://www.infopt.demon.co.uk/gayhist.htm

Rictor Norton maintains this site, which includes many of his essays, as well as discussions of the queer canon and the "great queens" of history (e.g., Michelangelo, Sir Francis Bacon). Many of the essays on the site appeared in *Gay Roots,* Volumes 1 and 2 (1991, 1993). The information on the site is extensive and well-organized, and includes multiple links to other relevant sites. Norton is the author of several books, including *The Myth of the Modern Homosexual: Queer History and the Search for Cultural Unity* (1997) and *My Dear Boy: Gay Love Letters Through the Centuries* (1997).

Gerber Hart Library
http://www.gerberhart.org

The Gerber Hart Library was founded in 1981 and describes itself as "the Midwest's largest gay, lesbian, and bisexual circulating library, archives, and resource center." Part of the library's mission is to offer a wide range of cultural and educational programs aimed at "dispelling homophobia through sharing knowledge." The site includes links to other libraries and archives with large collections of queer-related holdings.

Note: Use of the printed collection is available only to current Gerber Hart members. However, the links to other libraries may help you locate printed resources in your area, as well as online resources offered at other libraries.

The Human Sexuality Collection
http://rmc.library.cornell.edu/HSC

The Human Sexuality Collection at Cornell University documents historical shifts in the social construction of sexuality, with a focus on lesbian and gay history and the politics of pornography. The Web site is designed to give guidance to people asking LGBTQ history questions, who may or may not be looking for primary sources, and who may or may not understand everything about research using primary sources.

The Lesbian Herstory Archives
http://www.datalounge.net/network/pages/lha/welc.htm

The Lesbian Herstory Archives exist to "gather and preserve records of lesbian lives and activities so that future generations will have ready access to materials relevant to their lives." The collection, located in New York City, is the largest and oldest lesbian archives in the world. The site includes information on how to use the archives from a distance at: <http://www.datalounge.net/network/pages/lha/about/main.htm#distance>.

The Lesbian History Project
http://www-lib.usc.edu/~retter/main.html

The Lesbian History Project's site, based at the University of Southern California, includes links to books, articles, dissertations, and theses on lesbian history, as well as photographs and links to other online archives of lesbian history. Links to online resources about the histories of lesbians of color and lesbians in other countries are included.

The One Institute
http://www.usc.edu/isd/archives/oneigla

The One Institute, founded in 1952, describes itself as "an independent, not-for-profit California educational institution which houses the world's largest research library on gay, lesbian, bisexual, and transgendered heritage and concerns." The Institute houses over two million items, including 800 videos, and is located on the campus of the University of Southern California. The institute also publishes the *International Gay and Lesbian Review,* and abstracts of articles in the journal can be accessed online. The journal documents over half a century of publications on same-sex love.

Out of the Past: 400 Years of Gay History
http://www.pbs.org/outofthepast/home.html

This site continues the exploration begun by the film of the same name, and offers information on nearly 400 years of lesbian and gay history. The site's online resources include an extensive bibliography of books on lesbian and gay history, and links to other history-related sites.

Reclaiming History
http://www.uic.edu:80/depts/quic/history/reclaiming_history. html

This page, hosted by the History Department at the University of Illinois at Chicago, features biographies of a number of famous LGBTQ writers, including May Sarton, Emily Dickinson, James Baldwin, Willa Cather, Audre Lorde, and Francis Bacon.

The World History of Male Love
http://www.androphile.org

The site provides a good array of resources for studying the history of male-male love across many cultures and time periods. It includes essays on sub-Saharan Africa, China, Renaissance Italy, pre-Columbian America, Classical Greece, Native America, Oceania, Japan, Northern Europe, and Arab lands. You can use the site in a variety of languages, including English, German, and French. The site includes an online library featuring poetry, biographies, articles on myths and folk tales, anthropology, and history. This is an excellent site for those interested in history prior to the eighteenth century.

ETHNIC STUDIES

Bint el Nas
http://www.bintelnas.org

Funded in part by the Horizons Foundation, sponsored by the Queer Cultural Center, and produced by the Mujadarra Girls, Bint el Nas is a site designed for queer women who identify ethnicially or culturally with the Arab world, regardless of where they live. Bint el Nas contains essays, writings, and artwork, giving viewers a personal view of queer Arab women's experiences.

BLK Homie Pages
http://www.blk.com/blkhome.htm

The BLK Homie pages "provides news and information for black people in the life." "The life" refers to LGBTQ/same-gender-loving individuals. The site includes U. S. national listings of support groups and organizations and links to other online resources for black LGBTQ people. The site is the online home of the BLK Publishing Company, which publishes *BLK* and other magazines for black people in the life.

The Black List
http://www.blackstripe.com/blacklist

Professor Chuck Tarver at the University of Delaware compiled this site, which provides an alphabetical list of gay, lesbian, bisexual, and transgendered people of African descent, including a brief biography of each person listed. The site describes itself as "the Internet's leading resource for news, information and culture affecting lesbian, gay, bisexual, and transgendered people of African descent," and includes an extensive list of links relevant to queer people of African descent.

Gay and Lesbian Arabic Society
http://www.glas.org

The Gay and Lesbian Arabic Society (GLAS) is "an international organization established in 1988 in the U.S. to serve as a network for gay and lesbians of Arab descent and those living in Arab countries." The GLAS Web site is a good resource for research on queer Arab-Americans. The site includes a link to an Arab lesbian Web site that (as of June 2001) does not yet have content but promises to shortly at <http://www.glas.org/lazeeza.html>.

Lesbian History Project/Lesbians of Color
http://www-lib.usc.edu/~retter/loc.html

The Lesbians of Color site includes a wealth of publications, articles, interviews, bibliographies, a list of notable lesbian leaders of color, a chronology list, and links to other groups and sites.

LLEGÓ: National Latina/o Lesbian, Gay, Bisexual, and Transgender Organization
http://www.llego.org

LLEGÓ has worked effectively since 1987 to "build and strengthen the national network of Latina/o LGBTQ community-based organizations and to build their capacity to serve their local constituencies." LLEGÓ produces a variety of reports that are available online, and offers information and other resources about the LGBTQ Latina/o community.

Trikone
http://www.trikone.org

Trikone is a nonprofit organization for lesbian, gay, bisexual, and transgendered people of South Asian descent. Trikone was founded in 1986 in the San Francisco Bay Area, and claims to be the oldest group of its kind in the world. The Trikone Web site provides a unique directory of worldwide listings, publications, and organizations related to queer South Asians.

Works on Gay, Lesbian, Bisexual, and Transgendered Chicanos/as and Latinos/as
http://wwwvms.utexas.edu/~demedina/biblio.htm

Dennis Medina, who writes for *La Voz de Esperanza* and is a master's student at the University of Texas at Austin, maintains this useful resource on queer Chicanos/as and Latinos/as. The site includes an extensive list of books, chapters, and essays on LGBTQ Chicano/a and Latino/a issues.

RELIGIOUS STUDIES

Few topics generate more controversy and debate than the interplay between religion and homosexuality. This section focuses on online resources from Buddhist, Christian, Jewish, and Muslim sources, as well as Radical Faerie and Wiccan perspectives. At the time this guide was prepared, we were unable to find queer-focused online resources for a number of religions/spiritual beliefs.

Nondenominational LGBTQ Religious and Spirituality Resources

Partner's Task Force for Gay and Lesbian Couples
http://www.buddybuddy.com

Essentially a portal about same-sex partnering, this site provides more than 200 nuggets of information. It displays current news items one might find under discussion in *The Advocate*, as well as other topics that span a wide range. Readers will find a history of same-sex marriage legal battles and essays debating the issue, ample information on domestic partnering, and even information on where to get a religious blessing for a same-sex union (http://www.buddybuddy.com/blessing.html). The site provides a list of religious denominations that have commitment ceremonies for same-sex couples. It also features links to a few religious organizations that affirm such commitment ceremonies.

Religious Tolerance
http://www.religioustolerance.org

One might proclaim "Eureka!" when stumbling upon this site—it is like striking database gold when conducting queer-related religious research. The site promotes tolerance toward all faiths, from Christianity to Wicca. It also gives visitors many sides of various issues, such as homosexuality (http://www.religioustolerance.org/homosexu.htm) and how certain faiths view homosexuality. Religious Tolerance is among the most visited religious sites on the Web, receiving more than 7 million hits each month.

Sisters of Perpetual Indulgence
http://www.thesisters.org

Founded in the Bay Area in 1979, the Sisters of Perpetual Indulgence is now an international organization with several local chapters. The Sisters are an order of nuns who take vows to "promulgate universal joy and to expiate stigmatic guilt". Like tradtional nuns, their vows reflect the organization's commitment to their community. In San Francisco, the Sisters are involved in several community fundraisers, giving grants to the community via queer nonprofits and other agencies. Although there is quite a lengthy process to become a Sister of Perpetual Indulgence, the order welcomes men and women who identify as gay, heterosexual, bisexual and transgendered. The site includes links and a list of frequently asked questions about the mission and goals of the Sisters.

Soulforce
http://www.soulforce.org

Soulforce is an LGTBQ spiritual organization founded by Mel White. With a mission of "seeking justice for God's lesbian, gay, bisexual, and transgendered children," Soulforce frequently makes the news for its nonviolent protests against homophobic religious organizations. The Soulforce Web site includes news about current issues such as sodomy laws, same-sex marriage, employment, adoption, and hate crimes. The Evidence section includes articles and links providing scientific, psychiatric, biblical, and medical evidence related to controversial issues such as the origins of sexual orientation. The site also explains the Soulforce "journey" and vows.

White Crane Journal
http://www.whitecranejournal.com/wc05000.htm

The "Other Gay Spirituality Sites" section of the *White Crane Journal* provides an extensive list of links to online religious and spirituality sites, including queer nature spirituality, Will Roscoe's Web site (Roscoe is called the "father of gay spirituality" by many, and is the author of several books, including *Queer Spirits* and *The Zuni Man-Woman*), Radical Faerie online resources, Nasalam (a polysexual, polyamorous spiritual community), gay and lesbian Buddhist sites, EroSpirit (the Web site of Joseph Kramer, founder of the Body Electric School), gay Tantra sites, and other spirituality resources. This is an excellent starting point for assessing the types of online resources that are available on LGBTQ spirituality.

Buddhism

Gay Buddhist Fellowship
http://www.gaybuddhist.org

The Gay Buddhist Fellowship offers a monthly newsletter on topics of interest to LGBTQ Buddhists (however, it primarily focuses on gay male issues). You can access current and past editions of the newsletter on the site, which also includes links to other queer Buddhist online resources, and to sites that provide general information about Buddhism.

Christianity

Affirmation (Methodist)
http://www.umaffirm.org

Affirmation is an organization devoted to the concerns of gay, lesbian, and bisexual Methodists. The site provides links to several news items on the first page. For views on the United Methodists Church's stance on homosexuality, refer to the Religious Tolerance Web site (described previously) at <http://www.religioustolerance.org/hom_umc.htm>.

Affirmation (Mormon)
http://www.affirmation.org

Affirmation is a site for Mormons seeking to bring together their sexuality and their faith. A fairly extensive site, Affirmation has personal ads, electronic mailing lists for women, teen discussion forums, and media resources.

Catholic Lesbians
http://www.catholiclesbians.org

This site promotes the visibility and community-building of Catholic lesbians. Visitors may peruse the site's pastoral resources and its archived material. The archives contain new submissions and some reprinted articles from the organization's quarterly newsletter. Topics range from spirituality to sexuality, and from family to lesbian rights.

Catholic Theology Library
http://www.mcgill.pvt.k12.al.us/jerryd/cathmob.htm

This site contains a large amount of information. Links to a great many Catholic Church documents or essays on the official position of the church can be found on this site, from Vatican II to the Church's view of sexual morality. This is not a site that targets queer Catholics, but it would definitely aid in the research of someone interested in queers and the Catholic Church. It also provides links to a few Catholic search engines, and contains links to information on many forms of non-Catholic Christianity.

Dignity USA
http://www.dignityusa.org

Dignity USA is a national organization with local chapters in many cities. Dignity works with LGBTQ Roman Catholics to show that their sexuality and spirituality can be successfully integrated. The site contains several news items that detail events that affect queer Catholics. There are three mailing lists to which users may subscribe, including one for transgendered Catholics.

Integrity USA
http://www.integrityusa.org

Integrity calls itself "the leading grassroots voice for the full inclusion of homosexual persons in the Episcopal Church and our equal access to its rites." The nonprofit organization is more than twenty-five years old, and its site provides a wealth of information for the queer Episcopalian. Highlights include official church resolutions from the church's annual general convention, and a page with links to other LGBTQ faith organizations and antigay Episcopal sites.

Lutherans Concerned—North America
http://www.lcna.org

For information relating to LGBTQ concerns within the Evangelical Lutheran Church in America, visit the home of Lutherans Concerned—North America. This site is not as comprehensive as other

LGBTQ religious sites. Rather, the organization is a "pastoral care" group working with queer Lutherans to find an inclusive home within the Lutheran Church. The site includes a list of links to help you find an inclusive Lutheran church near you.

Metropolitan Community Church
http://www.ufmcc.com/menu.htm

The Metropolitan Community Church (MCC) has more than 300 congregations and 40,000 members. Started in 1968 by an openly gay clergyman, Reverend Troy D. Perry, MCC is the church of choice for many LGBTQ individuals who do not feel fully affirmed by the Christian faith they grew up with. The site includes a number of articles and essays about the Bible and homosexuality, including the following titles: "Homosexuality and the Bible: Bad News? Or Good News?", "Our Story Too: Lesbians and Gay Men in the Bible", "HIV/AIDS . . . Is It God's Judgment?", and "Homosexuality: Not a Sin . . . Not a sickness."

More Light
http://www.mlp.org

The More Light Presbyterians push for full inclusion and participation of LGBTQ people within the Presbyterian Church (USA). Highlights include a large resource page and updated news items of interest to Presbyterians or those studying them.

Reconciling Congregation Program
http://www.rcp.org

The Reconciling Congregation Program (RCP) describes itself as "a growing movement of United Methodist individuals, congregations, campus ministries, and other groups which publicly welcome all persons, regardless of sexual orientation." The site includes access to some of the articles printed in RCP's magazine, *Open Hands*, and to links of interest to reconciling congregations.

Religion and Homosexuality: The Lesbian, Gay, and Bisexual Catholic Handbook
http://www.bway.net/~Ehalsall/lgbh/#c3

Don't let the title of this site fool you—it is much more than a Catholic handbook. If you want to find out the truth behind the biblical quote "Thou shalt not lie with man as with a woman," or want to arm yourself against the fundamentalist homophobia, check out this site. It provides links to essays on specific biblical passages that are often used to condemn homosexuality, giving readers several viewpoints from which to argue the meaning or interpretation of the passages.

The site also includes an extensive bibliography, which is maintained by Paul Halsall at Fordham University. In addition to Catholic-related topics, you will find links to non-Christian spirituality sites, international queer religious sites, an extensive list of links to essays on same-sex marriage, and a number of essays describing how to deal with religious attacks based on sexual orientation or identity. The extensive online bibliography is an excellent resource for anyone doing research on religion and homosexuality.

St. Mark's Cathedral—Seattle
http://www.saintmarks.org

Visit this site for a view of an Episcopal church run by the highest ranking gay Episcopalian priest—the Reverend Robert Taylor. He's the first openly gay dean, or head, of an Episcopal Church. The site offers women's spirituality groups, a homeless assistance program, a poetry group, and Bible study and adult book groups. The site includes the public announcement of Taylor's call to be the sixth dean of the church, describing St. Mark's congregation's decision to have an openly gay man as its leader (http://www.saintmarks.org/DeanAnnounce.htm).

TransFaith
http://www.angelfire.com/on/otherwise/transfaith.html

This site caters to individuals who are members of both a Christian and a trans community—"transexuals, intersexuals, cross dressers, transvestites, and all other transgendered individuals, however they may be defined." Many articles on related topics are archived on this site, along with a lengthy glossary of trans and gender-variant terms.

Whosoever: An Online Magazine for Gay, Lesbian, Bisexual, and Transgendered Christians
http://www.whosoever.org

Whosoever's mission is to "break the bonds of extremist tyranny and rescue the Bible from fundamentalists who use the scripture to exclude and injure any of God's children." The site is a great resource for LGBTQ Christians who want to embrace their Christian faith rather than turn or run from it, while affirming their own or another person's sexual identity.

Judaism

Twice Blessed: The Jewish GLBT Archives Online
http://www.usc.edu/isd/archives/oneigla/tb

This site contains a fairly comprehensive assortment of information about queer Jewish issues—from artistic to historical to political. It is another good site from which to start research. It also contains a global listing of queer Jewish congregations and organizations. The site is also trans-inclusive.

World Congress of Gay, Lesbian, Bisexual, and Transgender Jewish Organizations Member Organizations
http://www.wcgljo.org

This site provides a comprehensive listing of LGBTQ Jewish congregations, queer-welcoming congregations, and organizations from around the globe. It gives a brief background on the World Congress of Gay, Lesbian, Bisexual, and Transgender Jewish Organizations (WCGLBTJO), plus information on upcoming conferences. The site is a good jumping-off point for people in search of local information regarding queers and Judaism. It also contains one current and one previous issue of the WCGLBTJO's quarterly newsletter *World Congress Digest*.

Muslim

Al-Fatiha Foundation
http://www.al-fatiha.net

Appropriately named (Al-Fatiha means "the opening"), the Al-Fatiha Foundation exists for queer and questioning Muslims and their friends. The organization began as an electronic mailing list before growing into its current structure as an international organization that hosts both international and North American retreats in addition to providing other services. The site provides links to news items about queer Muslims in the general queer media, plus links to several other sites, books, and related articles.

Queer Jihad
http://www.stormpages.com/newreligion/index.html

Queer Jihad, created by Islam convert Sulayman X, functions as a starting point for research on queer Muslims and views of homosexuality within the Islamic religion. Its name represents a queer Muslim's struggle with sexuality; "jihad" means either the struggle with oneself or a holy war. Adding welcome personal touches, Sulayman accepts written contributions related to being queer and Muslim.

Radical Faeries

Gay Guides Listing of Radical Faeries and Gay Spirituality Sites
http://www.gayguides.com/jeremy/faerielinks.html

This page provides links to Radical Faerie groups in the United States and Europe, and information about Faerie gathering places, or sanctuaries. Although these sites generally only provide information about planned events and individual groups, a review of these sites will give you a sense of what the Radical Faerie approach to spirituality and life is about.

Nomenus
http://www.nomenus.org

Nomenus is a "non-profit religious organization founded in 1984 to create, preserve, and manage places of spiritual and cultural sanctuary, for Radical Faeries and their friends to gather in harmony with nature, for renewal, growth, and shared learning." Nomenus maintains the Radical Faerie Sanctuary in Wolf Creek, Oregon. The site includes numerous links, information about various projects (e.g., the Harry Hay Documentary project), and other useful resources for those interested in learning more about the Radical Faerie movement.

RFD Magazine
http://www.rfdmag.org

RFD magazine is a "country journal for gay men everywhere." *RFD* is an online version of the print magazine that contains articles by Radical Faeries and other gay men living a rural lifestyle. The site lists the contents of the current issue of *RFD* and features a listing of Radical Faerie groups throughout the world. It also links to the Rural Gay Web site (http://www.ruralgay.com), which offers online articles and other resources, some of which focus on Radical Faerie issues.

Wicca/Pagan Spirituality

Gay Pagans, Gay Witches...Gay Witchcraft
http://www.thewellhead.org.uk/GP/gay1.htm

This site provides answers to questions about gay witches and the "Craft," including "Is there a place for gay men or lesbians in the Craft?" "Can I be a witch if I'm gay?" and "Does gender or sexuality really matter to the Gods?" Numerous links to queer- and non-queer-related pagan and witchcraft sites are also included.

Queer Dimensions
http://www.angelfire.com/pq/queerdimensions/

Queer Dimensions is a site for LGBTQ pagans. It includes links to queer pagan journals and events. It also includes links to guides on a variety of topics, including pagan terminology, near-death experiences, spirit guides, and ghosts.

Right-Wing Christian Hate Groups

God Hates Fags
http://www.godhatesfags.com

Yes, this is an actual site. It belongs to the Westboro Baptist Church, home of Reverend Fred Phelps, the church's pastor. This site is a good place to encounter viewpoints opposed to those espoused by Christian churches that support or accept LGBTQ people. Godhatesfags.com even provides a link to the Human Rights Campaign's listing of companies that mention LGBTQ employees in their nondiscrimination clauses and that offer domestic partner benefits, so that those who support Phelps and his ilk can avoid spending their cash on the products and services offered by the companies.

Information from other right-wing Christian hate groups can be found at the Traditional Values Coalition site (http://traditionalvalues.org), and the American Family Association site (http://www.afa.net).

NON-WEB RESOURCES

Gay Christians
http://groups.yahoo.com/subscribe/gaychristian

The Gay Christians mailing list functions as an online support group. It seeks to foster discussions such as how one may come out and how the Christian faith and homosexuality interrelate.

Gay, Lesbian, and Bisexual People of Color
majordomo@abacus.oxy.edu

This is a general discussion list focusing on the issues of race and sexual orientation. This list can help you keep current on issues affecting queers of color and provides a forum for discussion of work related to these issues.

Gay Muslims
http://www.al-fatiha.net/mlists.html

Gay Muslims is a mailing list for LGBTQ and questioning Muslims. The purpose of this list is to bring LGBTQ Muslims together to communicate issues of common concern.

Iman
http://www.queernet.org/lists/iman.html

Iman exists as a mailing list for lesbian, bisexual, and transgender Muslim women. It excludes biological males so that issues relevant to Muslim women may be discussed in a safe space.

Muslim Gay Men
http://www.OneList.com/subscribe/MuslimGayMen

This lists serves Muslim men who wish to remain true to their faith while simultaneously growing to accept and understand their (nonhetero) sexuality. Discussions include Islamic issues of primary concern such as al-Qur'an, Sunnah, Hadith, culture and family, and how to reconcile faith with social and religious rejection.

Queer Arabs
http://leb.net/glas/qa.html

The Queer Arabs mailing list is a forum in which "Arab and Middle Eastern-identified queers can discuss queer Arab culture, experiences, and issues." Friends, family, allies, and significant others of queer Arabs are invited to participate.

Chapter 9

Social and Biological Sciences

Alan Ellis

Social scientific research covers the gamut of human behavior, and virtually anything that humans do may be the focus of research in one or more of the disciplines found in the social sciences. These disciplines include anthropology, political science, psychology, and sociology. The following resources include the Web sites of the largest associations in each of the disciplines—each of which includes a search function that accesses LGBTQ-related materials. The biological sciences are included in this section largely because there are few online resources directly related to LGBTQ concerns and the hard sciences, and because the primary debate in the hard sciences focuses on the biological, genetic, and hereditary aspects of homosexuality—a debate that many social scientists also engage in.

Note: Citations in this section follow the American Psychological Association's standards, as APA style is used by the majority of the social sciences. A reference to this style is available at: <http://www.apa.org/journals/webref.html>.

GENERAL SITES

Dr. Greg Herek's Web site
http://psychology.ucdavis.edu/rainbow

For researchers in the social sciences interested in gay, lesbian, bisexual, and transgender topics, Professor Greg Herek's site at the University of California at Davis is perhaps the most useful resource on the 'Net. You can download current articles on a variety of research topics and link to multiple online resources that are categorized as follows: science and professional; polls, surveys, and other research studies; hate crimes and antigay violence; HIV/AIDS; policy and law; education and activism; other resources and organizations; and the religious right and related antigay links.

Regardless of the topic you're considering, spend some time exploring this site and the resources it offers.

The National Gay and Lesbian Task Force's Policy Institute
http://www.ngltf.org/pi

NGLTF's Policy Institute is "a proactive hub of research, policy analysis, tactical thinking, and strategic initiatives." Here you will find articles, fact sheets, and other resources—all available online. The Policy Institute has an Online Think Tank, which they describe as "the national research hub of accurate and cutting edge research, analysis, studies, reports, issues, facts, social science research, views, and tactical thinking." The Online Think Tank "provides the best research on GLBT issues to advocates, journalists, and researchers from the National Gay and Lesbian Task Force, colleague organizations, GLBT movement senior strategists, and academics."[1]

ANTHROPOLOGY

American Anthropological Association
http://www.aaanet.org

The site for the American Anthropological Association (AAA) lists a variety of section/interest groups within the association, including the Society of Lesbian and Gay Anthropologists (SOLGA), which is described as follows. AAA offers a list of an extensive and up-to-date Internet resources for anthropologists, some of which are specific to LBGTQ concerns (http://www.aaanet.org/resinet.htm).

Guide to Anthropological Resources on Sexuality Issues
http://www.library.adelaide.edu.au/guide/soc/anthro/subj/sex.html

This site, hosted by Adelaide University in Australia, offers an extensive bibliography on anthropology and sexuality divided into various categories, including bisexuality, homosexuality, transgender issues, and sexuality.

The Society of Lesbian and Gay Anthropologists
http://www.usc.edu/isd/archives/oneigla/solga/

The Society of Lesbian and Gay Anthropologists (SOLGA) "promotes communication, encourages research, develops teaching materials, and serves the interests of gay and lesbian anthropologists within the association."[2] The site includes articles on the history of gay and lesbian issues in anthropology and of SOLGA, and offers links to other useful sites.

POLITICAL SCIENCE

The American Political Science Association
http://www.apsanet.org

The American Political Science Association (APSA) is the largest U.S. organization for researchers in political science. The APSA Web site offers a very useful search function. For example, entering "gay," "lesbian," "bisexual," or "transgender" will bring up a list of online research articles and other resources.

Center for Lesbian and Gay Studies
http://web.gsuc.cuny.edu/clags/home.htm

The Center for Lesbian and Gay Studies (CLAGS), located at the City University of New York is "the first and only university-based research center in the United States dedicated to the study of historical, cultural, and political issues of vital concern to lesbian, gay, bisexual, and transgendered individuals."[3] The site includes information regarding CLAGS publications and numerous links to other helpful online resources.

Gay/Lesbian Politics and Law: WWW and Internet Resources
http://www.indiana.edu/~glbtpol

This site is described as an annotated guide to multiple resources on politics, law, and policy. It is maintained by Steve Sanders at Indiana University and is an excellent resource for those doing research on LGBTQ politics and law.

International Lesbian and Gay Association
http://www.ilga.org

The International Lesbian and Gay Association (ILGA), founded in 1978, is a "world-wide federation of national and local groups dedicated to achieving equal rights for lesbians, gay men, bisexuals and transgendered people everywhere." It is comprised of more than 350 member organizations. ILGA's Web site provides information about the organization and its current activities, including press releases and urgent action campaigns. It includes an extensive section of international news and the "World Legal Survey" of the status of LGBTQ rights around the world. Unfortunately, the site has not been updated in many months and should thus be regarded as a source of recent historical rather than cutting-edge information.

International Gay and Lesbian Human Rights Commission
http://www.iglhrc.org

The mission of IGLHRC is to "protect and advance the human rights of all people and communities subject to discrimination or abuse on the basis of sexual orientation, gender identity, or HIV status." Its Web site is a good source of international LGBTQ news, especially in the area of human rights. The World Watch section (http://www.iglhrc.org/world/index.html) provides a clickable world map with information about the LGBTQ human rights situation and current campaigns in different regions and countries. The site also features frequently updated action alerts organized by country, articles and press releases about IGLHRC's work, online versions of many of the group's books and fact sheets, and a good selection of links to resources related to human rights, immigration, political asylum, and international religions.

Institute for Gay and Lesbian Strategic Studies
http://www.iglss.org

The Institute for Gay and Lesbian Strategic Studies (IGLSS), described in Chapter 4, "Queer Studies," is particularly useful for research in political science. The site includes a "GayDAR" resource, which offers a directory of experts that journalists, policymakers, activists, and researchers can contact.

The Lesbian and Gay Caucus for Political Science
http://www.rci.rutgers.edu/~rbailey/caucus

The Lesbian and Gay Caucus for Political Science is the principal association of lesbians and gay men within the American Political Science Association. The site serves as a forum for the presentation of research on the interaction of sexual identity, theory, and political behavior.

Log Cabin Republicans
http://lcr.org

Log Cabin Republicans is the largest national organization for LGBTQ Republicans. Their site provides current news and information of interest to gay Republicans. Here you will find views of gay and lesbian concerns from a conservative perspective.

National Organization of Women: Lesbian Rights
http://www.now.org/issues/lgbi/lgbi.html

The National Organization of Women (NOW) provides an extensive collection of resources regarding political activism and the rights of lesbians and LGBTQ issues in general. Included are press releases, NOW policy statements, and information about the Lesbian Rights Summit held in 1999.

Gay and Lesbian Victory Fund
http://www.victoryfund.org

The Gay and Lesbian Victory Fund is a national organization committed to increasing the number of openly gay and lesbian public officials at all levels of government. Researchers can find information about current gay and lesbian candidates throughout the United States on the group's Web site.

PSYCHOLOGY

American Psychoanalytic Foundation
http://www.cyberpsych.org/apf/index.html

The American Psychoanalytic Foundation is dedicated to the advancement of psychoanalysis, and has as its mission "to educate the public, the community of mental health workers, and allied disciplines about the relevance of psychoanalysis as a powerful therapeutic and research instrument."[4] The Topics section of the site includes an essay on homosexuality and psychoanalysis.

American Psychological Society
http://www.psychologicalscience.org

The American Psychological Society's (APS) mission statement is "to promote, protect, and advance the interests of scientifically oriented psychology in research, application, and the improvement of human welfare."[5] The site has a search function, and entering the term "gay" returns over a dozen articles or research summaries.

The American Psychological Association
http://www.apa.org

The American Psychological Association (APA) has more than fifty divisions, including the Society for the Study of Social Issues (Division 9) and the Society for the Psychological Study of Lesbian, Gay, and Bisexual Issues (Division 44). The site includes a useful search function that accesses online articles on a variety of LGBTQ-related topics.

APA Division 44: The Society for the Psychological Study of Lesbian, Gay, and Bisexual Issues
http://www.apa.org/divisions/div44

APA's Division 44 is dedicated to the study of queer issues. The home page notes that "the Society for the Psychological Study of Lesbian, Gay, and Bisexual Issues was founded in 1985 as a Division of the American Psychological Association to represent sexual orientation issues within and beyond the Association" (as is the case with many organizations and associations, bisexual was added at a later date).[6] The division has more than 1,500 members who come from all fifty states, the District of Columbia, and eleven foreign countries.

The mission of Division 44 is to "strive to advance the contribution of psychological research in understanding lesbian, gay, and bisexual issues [in order] to promote the education of psychologists and the general public."

The home page of the division includes a Research link that leads to *Division 44 Annual* (a publication authored by the division and published by Sage Publications); the Joint Task Force on Professional Practice Guidelines, which works with the APA to establish guidelines for working with LGBTQ clients; the Public Policy Committee, which establishes the division's responses to public policy issues; the Science Committee, which maintains a directory of researchers on gay, lesbian, and bisexual issues; and member's publications (an alphabetical listing of recent publications by memmers of Division 44). The members' publication list is an excellent resource for identifying the most current areas of LGBTQ research in psychology.

APA Policy Statements on Lesbian and Gay Issues
http://www.apa.org/pi/statemen.html

The APA issues policy statements on topics of concern to psychologists, including public policy statements on LGBTQ issues. These statements address several issues including: discrimination against homosexuals; child custody and placement; employment rights of gay teachers; use of the diagnoses "'homosexuality'" and "'ego-dystonic homosexuality;'" hate crimes; AIDS and AIDS education; sodomy laws; Department of Defense policy on sexual orientation and advertising in APA publications; lesbian, gay and bisexual youths in the schools; and a resolution on state initiatives and referenda.

National Association for Research and Therapy of Homosexuality
http://www.narth.com

The National Association for Research and Therapy of Homosexuality (NARTH) is composed of psychologists and psychiatrists who oppose the American Psychiatric Association's and the American Psychological Association's determinations that homosexuality is *not* a mental illness. These individuals are dedicated to the "prevention" of homosexuality and to furthering societal condemnation of it. If you are researching the topic of reparative therapy (which both APAs have condemned), you might wish to check this site for information from such individuals as Paul Cameron, Joseph Nicolosi, and others who claim that homosexuality can be "cured." For an excellent critique of Cameron's work, see <http://psychology. ucdavis. edu/ Rainbow/html/facts_cameron.html>.

National Institutes of Mental Health
http://www.nimh.nih.gov

The National Institutes of Mental Health (NIMH) is a U.S. government agency that awards millions of dollars in funding for mental health issues, including AIDS prevention and education, and studies of teen suicide risk. Queer-specific topics can be found using the search function at this site, which is located on the pages linked to the home page. For example, click on "For the Public" and you will find the "Search" link in the left-hand column of that page. A recent search on the term "gay" brought up a number of articles and research projects, with the highest match being an article on "Frequently Asked Questions About Suicide."

SOCIOLOGY

The American Sociological Association
http://www.asanet.org/

The site for the American Sociological Association (ASA) lists a variety of section/interest groups within the association, including Sociology of Sexualities and Sex and Gender. In addition there is a caucus for LGBT sociologists. The caucus's URL and descripton are provided as follows.

Gay Wired Bibliography
http://www.gaywired.com/gaybooks/sociology.html

GayWired provides a bibliography of books focusing on sociology and the queer community and includes a brief description of most of the books listed.

The Sociologists' Lesbian, Gay, Bisexual, and Transgendered Caucus of ASA
http://www.qrd.org/qrd/www/orgs/slgc/SLGC.html

The Sociologists' Lesbian, Gay, Bisexual, and Transgendered Caucus consists primarily of members of ASA who have organized to "encourage unprejudiced sociological research on lesbians and gay men and their social institutions; provide a forum for current research, teaching methods and materials, and professional issues relevant to homosexuality; monitor anti-gay ideologies in the distribution of sociological knowledge and to investigate practices oppressive to lesbians and gay men; oppose discrimination against gay and lesbian sociologists in employment, promotion, tenure, and research situations; and maintain a social support network among its members."[7] The site offers a useful list of links for LGBTQ researchers in sociology, and an online version of the caucus's newsletter.

BIOLOGICAL SCIENCES

The primary focus of the biological sciences in the area of sexual orientation has recently centered on the essentialist argument that sexual orientation is either genetic or the result of hormone levels during a critical period of gestation in the womb. The countering argument—that sexual orientation is socially constructed—forces the debate to cross the line between the biological and the social sciences. As a result, some of the best online resources on this debate can be found at Dr. Greg Herek's Web site (discussed in the General Sites section of this chapter). In addition to information available online, useful print resources include *Gay Science: The Ethics of Sexual Orientation Research* (T. Murphy, Columbia University Press, 1997); *If You Seduce a Straight Person, Can You Make Them Gay?* (J. De Cecco and J. Elia, The Haworth Press, 1993); and the chapter on sexual orientation in *Psychological Perspectives on Human Sexuality* (A. Ellis and R. Mitchell, John Wiley and Sons, 2000).

Book Review and Article on the Biology of Sexual Orientation

Gay Science Book Review
http://www.arts.monash.edu.au/bioethics/murphy.htm

Here you will find a very good review by Udo Schüklenk—a lecturer at the Centre for Human Bioethics at Monash University in Melbourne, Australia—of the book, *Gay Science: The Ethics of Sexual Orientation Research* which appeared in the *Journal of the American Medical Association.*

Science, A Gay Gene, and the Third Sex
http://www.galha.freeserve.co.uk/glh163f3.htm

This URL provides access to an excellent article titled "Science, a Gay Gene, and the Third Sex" by Udo Schüklenk. The article argues that there is no evidence yet for a biological causation of homosexuality.

NON-WEB RESOURCES

Gay/Lesbian/Queer Social Science List
http://www.qrd.org/www/orgs/slgc/glqsocl.html

The Gay/Lesbian/Queer Social Science mailing list is hosted by the State University of New York at Binghamton.

Lesbian and Gay Caucus of the American Political Science Association
http://www.rci.rutgers.edu/~rbailey/caucus/LGB_APSA.htm

This is the mailing list for the Lesbian and Gay Caucus of the American Political Science Association.

Listing of Newsgroups in Anthropology
http://home.worldnet.fr/~clist/Anthro/Texts/usenet.html

This URL offers a listing of numerous Usenet newsgroups, including those focusing on social issues and anthropology.

INTERVIEW: ELLEN D. B. RIGGLE

Ellen D. B. Riggle, PhD, is Associate Professor of Political Science at the University of Kentucky. She is a co-editor (with Barry L. Tadlock) of *Gays and Lesbians in the Democratic Process* (Columbia, 1999) and (with Alan L. Ellis) of *Sexual Identity on the Job* (The Haworth Press, 1996). She has authored several articles on gay and lesbian issues, political tolerance, and evaluations of political candidates.

HMI*: What are some of the projects you are currently working on?

Ellen Riggle: I am currently working on a project with Professor Penny Miller of the University of Kentucky and Professor Ewa Golebiowska of Tufts University which examines newspaper coverage of openly LGBTQ candidates for political office. We are collecting data on about 45 year 2000 races throughout the United States to evaluate the type of coverage each candidate and their opponent(s) received in the local newspapers.

HMI: In what ways does the Internet play a role in your research?

Ellen Riggle: The Internet plays a big role in our research in many ways. We use the Internet to identify LGBTQ candidates through news stories and organizations that support the candidates. We are able to look at the current news stories and at earlier news coverage and stories by using the archives available for most of the local newspapers. We found a lot of newspapers using Google's directory of newspapers by state. We can also find information about candidates from their own campaign Web sites and from government Web sites (such as the Federal Election Commission, <http://www.fec.gov>; or state boards of elections). A lot of information that a researcher used to have to dig through libraries to find, or that simply was not available outside of the local area in which the candidate was running, is now available to people around the country through the Internet.

HMI: Are there Web sites or Internet resources that you use frequently?

Ellen Riggle: There are many Web sites that I use in the above-mentioned research and for other LGBTQ related projects. I use the Web sites of organizations that provide information relating to the issues I am researching, such as the Gay and Lesbian Victory Fund, the National Gay and Lesbian Task Force, and the Institute for Gay and Lesbian Strategic Studies. These organizations have press releases

---

*Interview conducted via e-mail, early November 2000, by Alan Ellis for HMI.

that are posted on their Web sites. The latter two also sponsor research, which can be downloaded from their Web sites.

A couple of other resources I use in order to locate relevant links are the Center for Lesbian and Gay Studies and the Lesbian, Gay and Bisexual Caucus for Political Science. The information on these Web sites and especially the links provided can be very helpful for finding information pertaining to political and policy issues.

Additionally, I use the news service of <fenceberry@aol.com>. The news service sends out several indexed e-mail messages every day with stories from newspapers around the country concerning LGBTQ issues. These articles often include links to the newspapers in which the stories appeared. I also find current news through the Web sites for *The Advocate* and PlanetOut.

HMI: What issues should a student or someone interested in conducting queer research in political science be aware of?

Ellen Riggle: It can be difficult to use search engines to find information specific to your topic. Keywords (such as "gay" or "lesbian") may or may not be useful. Either they may find too many matches to be helpful, or too few. Looking for links through Web sites which have already narrowed down your choices can help, such as the Queer Resources Directory (QRD). QRD is also a good source for information about international LGBTQ politics. For instance, the International Lesbian and Gay Association (ILGA) is listed there. The ILGA Web site includes information from organizations worldwide and can be very useful in studying the LGBTQ movement.

Students should carefully evaluate information available from Web sites. They must look carefully at the text and evaluate reported research thoughtfully. For instance, it is important to critically evaluate how the sample used in an opinion poll was selected. Online polls can be useful, but may not be reliable due to sampling method. One concern is that the Internet is primarily an English-speaking environment and is available primarily to those in higher socioeconomic groups. These factors can skew the results of an online poll.

When using the Internet, students should try to find information from a variety of perspectives. Often Web sites present a set of issues of broad concern but do not specifically represent the views or issues of LGBTQ persons of color. For instance, there needs to be a special effort made to seek political information on African-American or Latino/a gays and lesbians. For example, BLK Homie Pages (http://www.blk.com/blkhome.htm) lists organizations and publications for the black LGBTQ community. Also, little research has been done in political science specifically relating to transgendered persons. Students will need to look carefully for such information.

HMI: Is there anything else that you can think of that would help someone doing queer research in political science?

Ellen Riggle: The Web is a great resource for students. While they should not forget to look at primary sources that are not on the Web or to look to local groups and individuals for information, the amount of online information available on a worldwide basis concerning LGBTQ political issues is enormous.

NOTES

1. National Gay and Lesbian Task Force (no date listed). *NGLTF Policy Institute Description.* Washington DC: Retrieved January 19, 2001, from the World Wide Web: <http://www.ngltf.org/pi/thinktank.cfm>.

2. Society of Lesbian and Gay Anthropologists. (no date listed). *Organizational description.* Los Angeles: Department of Anthropology at the University of Southern California. Retrieved January 19, 2001, from the World Wide Web: <http://www.usc.edu/isd/archives/oneigla/solga>.

3. Center for Lesbian and Gay Studies, City University of New York. (27 September 2000). *Organizational description.* New York: Author. Retrieved January 19, 2001, from the World Wide Web: <http://web.gsuc.cuny.edu/clags/>.

4. American Psychoanalytic Foundation. (no date listed). *Organizational description.* Washington DC: Author. Retrieved January 19, 2001, from the World Wide Web: <http://www.cyberpsych.org/apf/index.html>.

5. American Psychological Society. (no date listed). *Mission statement.* Washington DC: Author. Retrieved January 19, 2001, from the World Wide Web: <http://www.psychologicalscience.org>.

6. The Society for the Psychological Study of Lesbian, Gay, and Bisexual Issues. (November 12, 2000). *Organizational description.* Washington, DC: Retrieved January 19, 2001, from the World Wide Web: http://www.apa.org/divisions/ div44/about_us.html mission.

7. The Sociologists' Lesbian, Gay, Bisexual, and Transgendered Caucus. (1999). *Organizational description.* St. Cloud, Minnesota: Retrieved January 19, 2001, from the World Wide Web: <http://www.qrd.org/www/orgs/slgc/SLGC.html>.

Chapter 10

Arts and Education

Alan Ellis

LGBTQ research in the arts often focuses on the specific works of LGBTQ artists and performers. Two excellent sites for determining current trends and activities in queer arts are Queer Arts Resource (QAR) and the Queer Cultural Center (Qcc). In addition, many of the Web sites in the history section of Chapter 8, "Liberal Arts and the Humanities" will be helpful in researching the lives of well-known LGBTQ artists and performers.

The key resource for those conducting research on educational issues for kindergarten through twelfth grade (K-12) relevant to the LGBTQ community is the Gay, Lesbian, and Straight Education Network (GLSEN). For those interested in higher education, the listing of programs—hosted at Duke University—is an excellent resource.

ARTS

Lesbians in the Visual Arts
http://www.lesbianarts.org

Dedicated to developing a dialoge about issues in contemporary art as well as issues specific to being a lesbian in the visual arts, Lesbians in the Visual Arts offers an online journal, *Pentimenta,* which includes interviews with artists and serves as the organization's newsletter.

Leslie Lohman Art Foundation
http://www.leslie-lohman.org

The Leslie Lohman Art Foundation showcases "patently gay art," primarily gay male art. An online archive of articles from the foundation's newsletters provides researchers with access to interviews with queer artists and other contemporary and historical information about art in the gay world. Fritz Lohman and Charles Leslie created the foundation in 1990.

Queer Arts Resource
http://www.queer-arts.org

Queer Arts Resource (QAR) provides an educational forum for the display and discussion of queer art and culture. According to QAR's mission statement, "until the recent advent of Queer Studies, the history of art has omitted most material of direct relevance to lesbians and gays. Much has been suppressed, much has been lost due to neglect or censorship, and a great deal has simply been overlooked. QAR is expanding the range and depth of knowledge about contemporary and historical queer art, and making this information freely available on our Web site."[1] (An interview with QAR's founder, Barry Harrison, can be found at the end of this chapter.)

On this site, you will find essays and ongoing discussions/comments regarding works of art. You will also find a large number of links to queer artists' Web sites, as well as other queer arts and culture links. This site is one of the most helpful sites on the Web for those interested in researching queer arts.

Queer Cultural Center
http://www.queerculturalcenter.org

The Queer Cultural Center (Qcc), the Center for Gay, Lesbian, Bi, and Transgender Art and Culture, conducts artistic and interpretive programs that explore queer identity issues. Qcc's Web site provides online access to contemporary and historical queer artists' work. The site includes a history of queer art and culture in San Francisco, starting with the 1889 visit by Oscar Wilde to the city. You can find information about the history of queer opera and the queer writers' project in the archives section. The site also provides links to numerous online resources that focus on artists such as Derek Jarman, Tee Corrine, and Jerome Caja.

EDUCATION

Gay, Lesbian, and Straight Education Network
http://www.glsen.org

The Gay, Lesbian, and Straight Education Network (GLSEN) describes itself as "the leading national organization fighting to end anti-gay bias in K-12 schools." GLSEN has forged alliances with a number of other significant national gay, lesbian, and mainstream organizations such as the National Education Association (NEA) to accomplish its goal. The site provides a large number of resources for those conducting research on LGBTQ issues in primary and secondary education, including links to hundreds of news articles on LGBTQ-related educational concerns and events. Overall, this is a gold mine for those interested in such issues as LGBTQ student safety, student athletics, and how to develop inclusive classrooms.

Lesbian, Gay, Bisexual, and Transgender Studies and Queer Studies Programs in Canada and the United States
http://www.duke.edu/web/jyounger/lgbprogs.html

This site, described previously in Chapter 4 "Queer Studies," provides a listing of university and college courses and programs offered throughout North America in LGBT and queer studies.

INTERVIEW: BARRY HARRISON

Barry Harrison received a master of architecture degree from the University of Pennsylvania and worked as an architect for ten years in Chicago and New York before moving to San Francisco. He founded QAR in 1996, and serves as its director.

HMI*: Barry, first let me say that you've designed a really beautiful site. Can you tell me how you personally use the Internet for scholarly research and/or business research?

Barry Harrison: Thank you. I wouldn't say I really do a lot of scholarly research, but I do spend a lot of time doing various types of research. I don't have any particularly sophisticated methods. I'm a

*Interview conducted by Elizabeth Taylor for HMI on May 25, 2000.

devotee of Yahoo, and I usually start a search there. Depending on the kind of search results I get, I try other search engines. Recently, I also started using the Northern Light search engine because of the convenient way that it categorizes things. It has all these different folders, and there are ways to narrow and define a search that sometimes leads to better results. The other way, of course, is to find a site that you find useful and pursue links from it. I think it's about fifty-fifty, whether you go to a search engine or just from one site to another.

HMI: Do you find searching for gay topics related to the arts any more difficult on the big sites like Yahoo? Do you use any particular gay sites?

Barry Harrison: I wouldn't use a large commercial site to search for gay subjects except to just begin a search. I would encourage people to go to the Queer Resource Directory or something that is really more focused.

HMI: What do you feel are the differences between Internet research and other kinds of research that are print-based?

Barry Harrison: One of the amazing things about the Internet is that you never know what you're going to find. The whole concept of hyperlinks is something that is so unique and so valuable. If you're willing, it can set you off in directions that you might not have anticipated when you started your search. I guess the same is probably true of books, but somehow it's just much easier if you're doing it online. When you're online you're clicking; you're not running from one library to another or one city to another trying to find resources. Those are some of the reasons that make using the Internet a preferable way to start searches.

HMI: While building the QAR site, what has been your biggest cyber influence? Is there a particular site or a particular person or group that you feel presented a starting point for you or a model?

Barry Harrison: The problem was that there were no good starting points, and there were no models for what we were trying to do. We launched QAR in 1996. It was a direct result of my first Websearching experience, in which a friend of mine showed me the Internet for the first time, and we did a search on Yahoo for gay art. There were a handful of responses that appeared. The first one was called "Men in Spandex." It was a site out of England; this gentleman had taken pictures from the *Undergear* catalog and posted that online as gay art. QAR was really a response to the lack of examples and models.

Other kinds of sites influenced me. For example, museum sites like that of the Museum of Modern Art. But back in 1996, there were very few of those types of sites, and what few there were weren't very fully developed yet. Museums were using their sites back then sort of as placeholders for listing business hours and membership information rather than actually having content or trying to do on-line versions of their exhibitions—and certainly not Web exhibitions.

The other art site that I want to mention in a general sense is Rhizome (http://rhizome.org). That was one of the first sites that really began to champion online art. It had a very sophisticated look and user interface and very active discussion groups. It was a really good way to learn about Internet-based art. It didn't have a specific focus towards gays and lesbians—although it was open to gay and lesbian art.

HMI: OK, this brings up another question about art and the digital information age. Does using the Internet as a means or a form of media for art work for you?

Barry Harrison: It works a lot better if you have a T1 connection! 'Net art is still in its infancy, although I do think it's sort of grown up a lot in the past couple years. Many artists don't know what to do with it. It's like giving people who have never seen a paintbrush and paints those tools and having them create. You're going to get a level of artistic achievement that's quite different than if they had been working with paints and brushes for twenty years. A lot of artists are innately skeptical of or downright scared of technology. But I have seen some Internet art which I found to be very powerful, moving, and strong.

HMI: On the QAR site, I've seen painting, sculpture, film, and video. Do you consider Internet art to be its own genre?

Barry Harrison: Yes.

HMI: In and of itself, there's really no other place for it but the Internet?

Barry Harrison: Right. I suppose you could put it on a CD, and view it on a monitor at a later date or offline. But you need the computer and the monitor to view it. I guess the other thing that's really so extraordinary about Internet art and really distinguishes it from any other kind of art is the notion of interactivity, and the fact that your experience of the piece is a direct result of your actions—your own clicking or following certain links.

HMI: In regard to QAR, what demographic information have you gathered? Who goes to the site? Is it predominantly people doing research? Art aficionados?

Barry Harrison: You know, unfortunately we've never had the resources to commission any kind of study about that, and it's not the kind of information that we just can pick up from looking at our Web logs. The only way I have to really answer that question is to think about it anecdotally. We have an online guest book and some discussion groups. In terms of what I see on those pages, I think that many people are looking for specific information; they're students doing reports. For example, "Was Andy Warhol gay?" You know, questions like that. They're looking for information about art and artists that has been suppressed by traditional sources for two thousand years. Even today, it's sometimes hard to find that information, particularly if you go to a library and primarily find books that are twenty years old.

In addition, I think another audience that uses QAR is artists who want to see what other people are doing and to learn about art from different perspectives. And finally, I think that there are people who are just curious, people who are interested in art and culture in general and want to learn more about it. And since we do combine both contemporary and historical artists on the site, it offers a good opportunity for people to get a broader sense of the contributions that gay and lesbian artists have made to our culture. So I would say that our audience mainly consists of those three primary groups.

HMI: How do you pick the art on your site?

Barry Harrison: We've developed a set of criteria that we use to evaluate potential exhibitors. We maintain very large lists of potential candidates, but since there are so many more artists than we actually have the resources to show, we try to create a set of criteria, which gives us a certain balance. We try to show both contemporary and historical figures, both male and female artists, and varied media. We are really trying to give as broad an overview as possible.

HMI: Is everyone gay that shows on the site?

Barry Harrison: That was never a question that we asked. We were really focusing on the content of the art. Yet, going back to the site's beginning, this was a big question. We had to think to ourselves, well, what is gay art? Is gay art a painting of a flower done by an artist who is identified as a lesbian? Or is gay art a photograph of two naked men frolicking on the beach? We decided that a better approach was to examine the content of the art.

One artist we exhibited is a straight woman; her name is Kimberly Austin. When she was a young girl, she had a cancer in her leg, so one of her legs was amputated. While all of her girlfriends were going out on dates and maturing, she was looking like a little boy. She was one-legged, and people were looking at her like they didn't quite know what she was. This meant that she really had to confront, in a lot of ways, the same kinds of identity issues that society forces gays and lesbians to confront all the time. And she uses that in her work; she uses a very interesting concept of an alphabet to define issues of gender and identity, which I thought was really directly related to our experience as queers, but in fact is not limited to that experience.

HMI: QAR is a not-for-profit organization. What is the economic viability of queer not-for-profit organizations on the Internet? Do you see them in any way related to bricks-and-mortar institutions dealing with the same types of issues and concerns?

Barry Harrison: Well, there are different kinds of not-for-profit organizations. QAR is unique because we weren't an existing not-for-profit organization that developed a Web site; we are entirely Web-based. That's problematic in that we don't have the resources, we don't have the membership, and we don't have the funding that most real-world organizations have. In fact, a lot of donors won't even consider a site like ours. Our situation is especially challenging, particularly in San Francisco; the city has the Hotel Tax fund, which funds just about every arts organization located in the city— except QAR.

The problem is that QAR is not located in the city; we're located on the Internet. If somebody were thinking about starting a not-for-profit organization, I wouldn't encourage him or her to do it on the Web in terms of funding. However, in terms of exposure and the opportunity to have a major influence with very little money, the Web has no equivalent. We spend a tiny fraction of what it would cost to operate a little neighborhood nonprofit gallery, and we have developed more than forty-six exhibitions that have been seen by hundreds of thousands of people.

HMI: How many people have visited the site?

Barry Harrison: I've never stopped to actually add it all up. We're getting a half a million hits a month, and we're getting almost one hundred thousand page views, but I don't know how that translates to actual individual people. That's really an entirely different technical calculation.

HMI: Are your viewers mostly in U.S.? Or are they international?

Barry Harrison: We've gotten hits from pretty much everywhere in the world— certainly mostly from the U.S. and other English-speaking countries. This makes sense because most of the content on the site is in English. However, we have made attempts to do Spanish translations and French translations of some of the exhibitions. We've also had hits from as far away as the United Arab Emirates and Pitcairn Island.

HMI: Has your audience changed? Are the people who were using QAR in 1997 or 1996 the same people who are using it now?

Barry Harrison: That's an easy one! I just don't know.

HMI: In terms of the people who actually produce the site and make the site work, how many are involved?

Barry Harrison: That has varied over time. Initially, the site was launched with just three of us: Jim Grady, Bert Green, and me. It has grown up to probably about a dozen people at its peak.

HMI: You've got an architectural degree from the University of Pennsylvania and were a practicing architect. So you came from a pretty creative and artistic profession in relation to informing the whole architecture of the site.

Barry Harrison: Yes. Right. I think design is design, wherever and whatever form it takes.

HMI: Are exhibitors picked from a long list by a jury?

Barry Harrison: We tried very hard to establish credibility in the beginning because when you start out, and you don't have really big names or big bucks behind you, all you can really do is try to find the best people that you can. So, one of the things that we found out quickly was that people who saw the site became very excited about it and really wanted to contribute. We had people coming out of the woodwork offering to do things. But we also were able to target certain individuals. For example, we planned an exhibition on David Wojnarowicz, but then we learned that the New Museum of Contemporary Art was going to do a real-world exhibition of his work. We contacted them and worked with their Senior Curator, Dan Cameron, to develop a Web exhibition in conjunction with the New Museum's exhibition. That's the kind of thing that can really give us more credibility because we can also talk about this relationship.

HMI: QAR has been written up in the *New York Times Online*, and there have been reviews in *The Advocate*, and *OUT* magazines. Are there any other recent publications or reviews?

Barry Harrison: My favorite was one of the earliest ones, which was an article in *Art in America*. It was titled "California Arts Organizations Brighten the Web." It listed the Fine Arts Museum of San Francisco Web site, the UCLA Web site, the Getty Institute Web site, and finally it listed Queer Arts Resource! I was just absolutely blown away that I was working in my little home office, our Webmaster was working in his bedroom, and QAR was on the same page as the Getty! That is the other thing about the Web; it doesn't matter that we were working out of a bedroom. From the user's perspective, or the visitor's perspective, it's irrelevant; all they see is the content. They can equally access the fabulous collections of the Getty and the Queer Arts Resource Web site.

HMI: Are there things about the Internet that you don't like? Things that you think are a disadvantage in relation to the mission and masses that you talked about? Education? Access? Are there factors you find limiting? Is there anything you would do differently?

Barry Harrison: I think part of the challenge for me as the director of QAR, and also just as a user of the Internet, is sifting through all the stuff that I'm just not interested in or that just isn't very good. On the Internet, there are no criteria for calling yourself an artist; anyone can do it, anyone can make a Web site. Do I want to see all of these artists' Web sites? No, very, very few. So the question is, "'How does a human being with tight real-time constraints deal with that kind of quantity of information?'" To me, that's really the big challenge, actually getting something that you're looking for that is useful and interesting.

HMI: How did you find the resources that you put on the Web site? You've got more than forty-six exhibitors. Did you find them online? Or was it through connections with people like curators?

Barry Harrison: I am trying to think if I ever found resources through artists online. Nothing comes to mind. I think it was all pretty much offline. Reading magazines, books, discussions, having the help of our members, volunteers, and board. There is never any shortage of names of artists. We occasionally get e-mail from people who want to have a show or a link to their site. Some of these are great, but the basic thought is, "Well, if they already have a site, it doesn't really make that much sense for QAR to do an exhibition on their work, since we can just set up a link to it and direct people to it."

HMI: You don't sell art on the site. Is there a reason for this? How does the site support itself?

Barry Harrison: Well, there is no reason set in stone as to why we don't sell art online. The biggest problem was that I found it very difficult to believe that anyone would want to buy art online. It seems like such a personal and particular kind of purchase that, speaking for myself, unless I knew already what I was buying (in other words, if I were already a collector of a certain person's work), it would be extremely difficult. There are just too many limitations. You're looking at something on a monitor. Even assuming that the color is perfect, the fact of the matter is you're having light coming from the piece rather than reflecting it as it would in the real world. And the image may be one-eighth the size of the real piece. So there just seems to be too big a disconnect between what you're seeing and what you're buying.

HMI: At the same time, we know that there's NextMonet.com. We know that there's Art.com. And they're in business.

Barry Harrison: Right. There are commercial art sites. I am very interested in watching them as they develop. I also wonder about potential conflicts between work you think sells and work you think should be seen but won't necessarily be purchased. Because we are really set up as an education and information resource; it's a different point of view. We could theoretically sell other kinds of art, other than that which we're exhibiting. We toyed with the idea of selling posters or certain reproduction pieces. But we didn't ever really develop that because, more from the technical point of view, it becomes kind of a big process.

HMI: Your site supports itself through volunteer work and your efforts?

Barry Harrison: We've received some grants. We've certainly received a number of donations from individuals. The Gill Foundation and the Horizons Foundation supported us. But the fact of the matter is that there aren't sufficient resources at this point to do anything other than maintain the site.

NOTE

1. *Queer Arts Resource*. No creation date listed. *Queer Arts Resource*. January 19, 2001. <http://www.queer-arts.org>.

Chapter 11

Law and Philosophy

Alan Ellis

The fight for gay, lesbian, bisexual, transgender, and queer rights and equality includes coming out individually and collectively, working toward supportive legislation, and winning in the courts. This chapter focuses on organizations that engage in litigation on behalf of the LGBTQ communities and that identify and summarize current legal concerns, any of which might be the basis for a research paper on queer legal issues. Also, see the Web sites and Internet resources listed in the political science section of Chapter 9, "Social and Biological Sciences," as well as the section on domestic partnerships and workplace nondiscrimination in Chapter 13, "Business, Labor Studies, and Economics."

LAW

Annual Review of Gender and Sexuality Law
http://www.law.georgetown.edu/journals/gender

The *Georgetown Journal of Gender and the Law* publishes an annual review of gender and sexuality law. This site provides information about the journal and about symposia and conferences on this topic.

Gay and Lesbian Advocates and Defenders
http://www.glad.org

Less well known than many of the other queer legal organizations (such as Lambda Legal Defense Fund and the Human Rights Campaign), Boston-based Gay and Lesbian Advocates and Defenders (GLAD) provides online articles, briefs, and discussions of the organization's current cases, including a variety of civil rights and AIDS law cases in the New England area.

Note: GLAD is a separate organization from the better known Gay and Lesbian Alliance Against Defamation (GLAAD).

Lambda Legal Defense and Education Fund
http://www.lambdalegal.org

Lambda Legal Defense and Education Fund describes itself as "the nation's [U.S.] oldest and largest legal organization working for the civil rights of lesbians, gay men, and people with HIV/AIDS."[1] Position papers and current news on legal and legislative cases affecting the queer community, as well as state-by-state issues and concerns, are included on the organization's Web site. In addition, you will find resources such as Lambda's Web-based Back-to-School Kit, which promotes safer schools for lesbian and gay students, fair employment conditions for LGBTQ teachers, and respect for all civil rights at school. Lambda's litigation is described in detail for each issue, and links to each case are provided. Recent cases have focused on AIDS, antidiscrimination, antigay initiatives, criminal law, domestic partnership, employment, family, first amendment rights, housing and public accommodations, immigration and political asylum, marriage, the military, older lesbians and gay men, reproductive rights, transgender issues, violence, and youth.

The site also includes a library of resources, arranged by cases, briefs, decisions, events, memos, news and views, press releases, publications, and resources.

Overall, this is an excellent online collection of resources for anyone conducting research on legal issues relevant to the queer community.

National Center for Lesbian Rights
http://www.nclrights.org

The National Center for Lesbian Rights' (NCLR) primary focus is on advancing the rights and safety of lesbians and their families. However, on key issues that significantly advance lesbian rights, NCLR offers support to gay, bisexual, and transgendered individuals as well.

NCLR's programs include free legal advice and counseling, public education, public policy advocacy, and litigation. The site includes information about current cases, events, projects, publications, and related links. For attorneys interested in taking part in this work, or being added to NCLR's attorney referral list, an online questionnaire is provided.

New England School of Law's Gay, Lesbian, Bisexual and Transgender Caucus
http://www.nesl.edu/students/outlaw.htm

Although this site is hosted by the New England School of Law's caucus, it provides a listing of law student caucuses throughout the United States, as well as links to resources for LGBTQ law students and those conducting legal research.

Transgender Law and Policy
http://www.trangenderlaw.org

This site, described in Chapter 6, "Transgender and Intersex Studies," features information and resources related to transgender legal and policy issues.

PHILOSOPHY

Much of the discussion and research in philosophy dealing with LGBTQ topics centers on issues of law and public policy. As a result, the Web sites listed in the law section of this chapter and the sites listed in the political science section of Chapter 9, "Social and Biological Sciences" will also be useful.

Erratic Impact
http://www.erraticimpact.com/~lgbt

Erratic Impact, a site that began as a research tool for students enrolled in a philosophy course at Villanova University, features "hundreds of annotated links to queer theory and LGBT resources, designed to assist lesbian, gay, bisexual, and transgendered people involved in academic study and philosophy research." The site is a good starting point for those conducting research in the discipline of philosophy on LGBTQ issues.

NOTE

1. Lambda Legal Defense Fund. No creation date listed. Lambda Legal Defense Fund. January 19, 2001. <http://www.lambdalegal.org>.

Chapter 12

Health and Medicine

Liz Highleyman

Health is one of the most well-covered areas on the Internet, with new sites cropping up every day. Information ranges from medical journals to sensational testimonials from the purveyors of alternative therapies. Much of the available information is good, but some is poor—even potentially dangerous. Users doing health-related research on the Web are encouraged to seek reputable sources and to verify the information they find. Although several Web sites focus on LGBTQ health, they are not as numerous or complete as one might hope, given the number of both non-health-related LGBTQ sites and nonqueer health sites.

GENERAL LGBTQ HEALTH ISSUES

Gay.com Health Channel
http://www.gay.com

Like most major Web portals, Gay.com features a health channel. The channel includes good sections on lesbian health, men's health, and gay sex. Breast cancer, hepatitis, and sexually transmitted diseases are well covered. Content includes LGBTQ health news, an "ask the doctor" feature, and discussion and chat areas. Gay.com also has an HIVlife and a sports and fitness channel, and a home and family channel that covers LGBTQ parenting issues.

Gay and Lesbian Medical Association
http://www.glma.org

The Gay and Lesbian Medical Association (GLMA) is a nonprofit organization of LGBTQ physicians, medical students, and patients working to end homophobia in health care. The association's Web site provides information about GLMA programs (such as the Lesbian Health Fund and physician referrals), news, legal and policy issues, upcoming conferences, and some patient-focused medical information. There is also a link to the *Journal of the Gay and Lesbian Medical Association,* featuring tables of contents and abstracts for selected articles.

GayHealth
http://www.gayhealth.com

This relatively new site is the primary Web resource dedicated solely to LGBTQ health. Comprehensive, well-designed, and easy to use, it has channels focused on sexual health, drug and addiction issues, mental and emotional health, image, food and fitness, and a society section with information about teens, seniors, violence, coming out, and legal issues. There is also a tool to help users locate queer health care providers and information for providers.

The GLBT Health Access Project
http://www.glbthealth.org

The GLBT Health Access Project is a collaboration between the Massachusetts Department of Public Health, Boston's Justice Resource Institute (JRI), and others, focused primarily on public policy related to GLBT health care. The site describes the project and its various programs and includes several useful reports, including *Health Concerns of the Gay, Lesbian, Bisexual and Transgender Community; Lesbian Health: Current Assessment and Directions for the Future;* and *Community Standards of Practice for Provision of Quality Health Care Services for Gay, Lesbian, Bisexual, and Transgendered Clients.*

McMaster University Health Sciences Library
http://www-hsl.mcmaster.ca/tomflem/gay.html

This site, put together by Tom Flemming of the McMaster University Health Sciences Library, is one of the best collections of health-related links on the Internet. The selected gay health links point to basic information about sexual orientation, coming out, gay men's health, lesbian health, and aging. Several of the links lead to local community health centers and projects. In addition to the gay health section, the site also includes collections of links related to gay and lesbian health care practitioners, gay and lesbian health problems, transgender issues, intersexuality, sexually transmitted diseases, sexual health, and adolescent sexuality.

ZapHealth's LGBT Health Issues
http://www.zaphealth.com/lgbt_comingout.htm

ZapHealth is a site devoted to providing health information for youth. Its LGBT Health Issues pages feature an article on coming out to your health care provider. It also discusses LGBTQ health-risk factors and provides a list of local queer-friendly health clinics. The main ZapHealth site (www.zaphealth.com) also includes sections on drugs, birth control, relationships, sexuality, sexually transmitted diseases, and violence.

GAY MEN'S HEALTH ISSUES

Gay Men's Health
http://www.gmhp.demon.co.uk

Offered by Gay Men's Health Wiltshire and Swindon in the UK, this site is a good Internet resource specifically focused on gay and bisexual men. It includes the full text of several guides produced by the project on topics including coming out, sexually transmitted diseases, and how to use condoms. It offers advice on how gay and bi men can better communicate with their health care providers. The site also provides a listing of useful Web links and a guide to UK community resources.

Gay Men's Health Summit
http://www.temenos.net/summit

The Gay Men's Health Summit was held in 1999 and 2000 with the aim of building a broad, multiissue gay men's health movement; summit organizers are working to develop local gay men's health projects and regional conferences in ensuing years. This Web site features conference schedules, press information, and the text of selected presentations from past summits. It includes a message board and information about mailing lists for those who want to get involved in organizing around gay men's health issues, as well as a growing list of links to relevant projects.

GayWellness
http://www.gaywellness.com

GayWellness is an attractice, well organized site devoted to the health issues of queer men. It features information about coming out, drug and alcohol problems, sexually transmitted diseases, safety and self-defense, and AIDS. The site also includes a database of providers and community resources, and an educational section for health care providers who have gay men as patients. A partner site, LesbianWellness, provides similar information for queer women.

Health Concerns Among Gay Men
http://www.thebody.com/sowadsky/gaymen.html

This site from The Body (see HIV/AIDS section, below) presents the full text of a report on gay men's health concerns written by Rick Sowadsky. The site interprets gay men's health broadly, looking beyond AIDS and sexually transmitted diseases to include cancer, alcohol and drug use, violence, and mental health.

LESBIAN HEALTH ISSUES

Lesbian.com Health and Wellness
http://www.lesbian.com/health/health_intro.html

Lesbian.com's health and wellness section features a good collection of nearly fifty links related to lesbian and women's health, herbal therapy, HIV/AIDS, spirituality, and more, including a starting point for the Lesbian Health Web Ring, a collection of related sites devoted to lesbian health.

Lesbian Health and Homophobia
http://www.ohanlan.com/lhr.htm

This site by Kate O'Hanlan, MD, presents a good report, *Lesbian Health and Homophobia: Perspectives for the Treating Obstetrician/Gynecologist,* that provides education for health care professionals who work with lesbians, including sections on various theories of sexual orientation diversity, homophobia, health conditions, families and relationships, the treatment of adolescent girls, and suggestions for improving health care for women who have sex with women.

Lesbian Health Links
http://www.lesbian.org/lesbian-moms/health.html

This site provides another extensive collection of links to a broad range of sites dealing with lesbian health and wellness, including sites dealing with homophobia, HIV/AIDS, and battered lesbians. The site is provided by the Lesbian Mothers Support Society. The group's home page, at <http://www.lesbian.org/lesbian-moms>, provides a plethora of resources for lesbian moms, including information about adoption, alternative fertilization, legal issues, and resources for children of lesbian parents.

LesbianWellness
http://www.lesbianwellness.com

LesbianWellness is an attractive, well organized site devoted to the health issues of queer women. It features information about coming out, cancer, safety and self-defense, and AIDS. The site also includes a database of providers and community resources, and an educational section for health care providers who have lesbians as patients. A partner site, GayWellness, provides similar information for queer men.

The Mautner Project for Lesbians with Cancer
http://www.mautnerproject.org

The Mautner Project presents an attractive, easy-to-use site that includes—in addition to services provided by the project—information about cancer risk factors and how to do a breast self-exam. The site includes information about breast cancer in Spanish, and a good list of other Web sites dealing with breast cancer and other cancers that affect women.

Pathfinder: Lesbians and Health Care
http://www.suba.com/~leskovec/lesbianhealth

This site offers an annotated bibliography of published and electronic resources on lesbian health care for both health professionals and lay people. In addition to a general overview of lesbian health care, the bibliography includes various reports and studies on woman-to-woman sexuality, lesbian identity, relationships, gynecology, and sexually transmitted diseases.

BISEXUAL HEALTH ISSUES

Bisexual Action on Sexual Health
http://bi.org/~kcl/bash.html

UK-based BASH is one of the few community health projects specifically by and for bisexuals. Its Web site provides the history of the organization, achievements, current projects, and contact information. There is also a small list of links to other bi sites and general (non-bispecific) health sites.

Bisexual Resource Center Health Section
http://www.biresource.org/health

The Bisexual Resource Center site presents one of the few listings of Internet health resources collected for bisexuals. Although most of the links are not bi-specific, this remains one of the best such listings, given the paucity of bisexual health sites elsewhere on the Web.

Playing Safe with Both Teams
http://www.biresource.org/bothteams

This site, sponsored jointly by the Bisexual Resource Center and the Fenway Community Health Center, includes the text of presentations from the "Playing Safe With Both Teams: Bisexuality and HIV Prevention" conference held in June 1999. Articles cover a range of topics from myths about bisexuality, to portrayals of bisexuals in the media, to building community programs. The site also has tips for health care providers working with bisexuals, and for bi patients on how to better communicate with health care professionals.

TRANSGENDER AND INTERSEX HEALTH ISSUES

Ingersoll Gender Center
http://www.ingersollcenter.org

Seattle's Ingersoll Gender Center is one of the best known gender dysphoria clinics in the United States. The Web site covers a range of topics of interest to both MTF and FTM transsexual and transgendered people. The health section collects information on important but non-transition-specific subjects such as nutrition, hepatitis, and other health concerns. The site also provides book reviews, skin care tips, and an article on breast exams for MTFs and FTMs.

Intersex Society of North America
http://www.isna.org

The Intersex Society of North America (ISNA) Web site described in Chapter 6, "Transgender and Intersex Studies" is the premier Internet resource for intersexed people. The "On Treatment" section includes articles about the treatment and medical management of intersexuality. The site gives examples of medical diagnoses associated with intersexuality: "clitoromegaly, micropenis, hypospadias, ambiguous genitals, early genital surgery, adrenal hyperplasia, Klinefelter Syndrome, androgen insensitivity, and testicular feminization."

Transgender Education Network
http://www.jri.org/ten.html

The Transgender Education Network (TEN), a project of Boston's JRI Health, is one of the best health-focused resources in the crowded field of Internet resources for transgendered people. TEN provides information, research, referrals, and trainings for transgendered people—both MTF and FTM—and their care providers. The Web site includes the Harry Benjamin International Gender Dysphoria Association (HBIGDA) standards of care for transsexuals, an explanation of gender-related terminology, a section on HIV/AIDS and substance abuse, and listings of Web links and New England resources for transgendered people.

Transsexual Women's Resources
http://www.annelawrence.com/twr

TWR is an attractive, comprehensive Web site for transsexual women put together by Anne Lawrence, MD, a practitioner of transgender medicine. The site includes detailed information on hormone therapy, sex reassignment surgery, postoperative care, cosmetic surgery, hair removal, and voice alteration, as well as Lawrence's *Journal of the American Medical Association* article "Health Care Needs of Transgendered Patients." There is also a good listing of Web resources, including links for transsexual youth and transsexual parents.

HIV/AIDS SITES

There may be more information on the Web about HIV/AIDS than any other disease. Because there are so many HIV/AIDS resources on the Internet, the following list is just a sampling of some of the largest, most established, and most comprehensive sites. Most of these sites contain news, research findings, prevention and treatment information, and much more.

AEGiS
http://www.aegis.org

AEGiS, AIDS Education Global Information System, is one of the oldest Internet resources on HIV/AIDS, long preceding the Web and dating back to the days of bulletin board systems in the mid-1980s. Updated hourly, AEGiS is one of the most comprehensive sites—so comprehensive, in fact, that it's easy to get lost among its reams of information; a good search engine helps users find what they're looking for. One of the site's unique features is its extensive archive of HIV/AIDS-related news from the queer and mainstream press. (An interview with the founder of AEGiS, Sister Mary Elizabeth, follows at the end of this chapter.)

The Body
http://www.thebody.com

The Body is a long-standing project focused on demystifying HIV/AIDS and its treatments and fostering community among people with HIV. Its comprehensive Web site covers medical, social, and quality-of-life aspects related to HIV/AIDS.

Gay Men's Health Crisis
http://www.gmhc.org

New York City's Gay Men's Health Crisis (GMHC) provides a useful Web site with a focus on prevention education. The site includes safer sex and safer drug use information, and access to full-text articles from GMHC's *Treatment Issues* magazine.

HIV Insite
http://hivinsite.ucsf.edu/InSite

This site from the University of California at San Francisco is among the most comprehensive HIV/AIDS Web sites, but it can be hard to navigate due to the vast amount of information available, ranging from medical to legal to social and policy information. Its strengths are research, treatment, statistics, and international issues. The Medical section includes the full text of *HIV Insite Knowledge Base* online textbook.

HIV/AIDS Treatment Information Service
http://www.hivatis.org

This U.S. government site, a project of the Department of Health and Human Services, provides regularly updated versions of the latest federal HIV/AIDS and opportunistic infection treatment guidelines for adults, children, and pregnant women.

HIVandHepatitis.com
http://www.hivandhepatitis.com

This site, run by a group of individuals involved in HIV and hepatitis education and activism, provides a comprehensive collection of news related to HIV/AIDS, with an emphasis on the latest treatment research findings. It also provides rapid and extensive coverage of relevant medical conferences.

Project Inform
http://www.projinf.org

Project Inform is one of the most well known community HIV/AIDS organizations, providing treatment-related education and advocacy. Its Web site provides links to Project Inform's vast library of fact sheets, treatment publications, and archives of *PI Perspective* magazine. There is also a section on HIV/AIDS treatment information for women and a Spanish-language section.

Rethinking AIDS
http://www.virusmyth.net/aids/index.htm

The Rethinking AIDS site presents news and information from the perspective of HIV dissidents who do not believe that HIV is the sole and sufficient cause of AIDS and who believe that anti-HIV drugs are harmful.

San Francisco AIDS Foundation
http://www.sfaf.org

The San Francisco AIDS Foundation (SFAF) is a large community-based organization providing treatment education, policy advocacy, and direct client support to people with HIV/AIDS. Its Web site is comprehensive, with good sections on prevention, safer sex and drug use, and public policy. Users can access full-text articles from SFAF's *Bulletin of Experimental Treatments for AIDS (BETA)* magazine and other publications. Information is also available in Spanish.

GENERAL CONSUMER HEALTH SITES

In addition to sites devoted specifically to LGBT health, many Web portals cover general health issues. Most of these sites include a combination of features such as libraries of diseases and conditions, medication and alternative therapy databases, interactive tools and calculators, locators for health care providers, "ask a doctor," and discussion and chat areas. As a rule, general consumer health sites do not include sections with queer-specific information, although they do often include health news items relevant to LGBTQ people, which can be found using keyword searches; try common words like "gay" or "lesbian" rather than community terms like "queer" or "GLBT." The following sites are some of the most comprehensive among the many such resources on the Web:

http://www.drkoop.com (Dr. Koop)
http://www.healthcentral.com (Health Central)
http://www.mayoclinic.com (Mayo Clinic)
http://thriveonline.oxygen.com (Oxygen Media's Thrive Online)
http://www.webmd.com (WebMD)

Other good resources, especially for research about a specific condition, are federal government Web sites and the sites of societies and organizations devoted to specific diseases. These sites are often the best source for comprehensive, in-depth information for people who are not medical professionals. You can find many of these sites by entering a disease or organization name into a search engines such as Google. The American Academy of Pediatrics provides a listing of the home pages of over sixty medical societies and nonprofit health organizations at <http://www.aap.org/advocacy/washing/orgresc.htm>.

The following sites provide just a sampling of the many resources available:

http://www.cancer.org (American Cancer Society)
http://www.americanheart.org (American Heart Association)
http://www.ama-assn.org (American Medical Association)
http://www.bcaction.org (Breast Cancer Action)
http://cancernet.nci.nih.gov (National Cancer Institute)
http://www.niaid.nih.gov (National Institute of Allergy and Infectious Diseases)
http://www.niehs.nih.gov (National Institute for Environmental Health)
http://www.nih.gov (National Institutes of Health)
http://www.cdc.gov (U.S. Centers for Disease Control and Prevention)
http://www.healthfinder.gov (U.S. Department of Health and Human Services HealthFinder)
http://www.fda.gov (U.S. Food and Drug Administration)

RESEARCHING THE MEDICAL LITERATURE

Perhaps the health researcher's best friend on the Internet is PubMed (http://www.ncbi.nlm.nih.gov/PubMed), a search service provided by the National Library of Medicine that allows users to access a vast compendium of articles in medical journals and other publications. PubMed is the gateway to MEDLINE, a database of over 11 million citations, as well as AIDSLINE, CANCERLIT, and other medical literature databases.

In addition, several medical journals also provide tables of contents, abstracts, and full-text articles on their own Web sites, for example *The Lancet* (http://www.thelancet.com), the *Journal of the American Medical Association* (http://www.jama.ama-assn.org), and the *New England Journal of Medicine* (http://www.nejm.org).

NON-WEB RESOURCES

Mailing Lists

There are more newsgroups, mailing lists, chat forums, and other non-Web resources on health topics than can be covered here.

There are many electronic mailing lists for people dealing with specific diseases and conditions. Often Web sites devoted to the condition in question will provide a link to a mailing list. Several of the large consumer health Web sites (such as WebMD) have their own discussion and chat forums, typically listed in a section called "Community" or something similar. Another way to find discussion groups of interest is to do a search on the Yahoo Groups Web site (http://groups.yahoo.com). Most health-related lists are in the Health and Wellness section of the directory. Search engines can also help you find relevant mailing lists and chat forums.

Usenet Newsgroups

Among Usenet newsgroups, many health-related forums fall within the **sci.med** hierarchy. The top-level **sci.med** newsgroup deals with all matters related to various medical conditions. Several common diseases have their own subsidiary newsgroups, for example **sci.med.aids.** Newsgroups that provide support and resources for people with specific conditions fall in the **soc.support** hierarchy, for example **soc.support.depression.** The

misc hierarchy also contains several health-related newsgroups, such as **misc.health** and **misc.fitness.**

INTERVIEW: SISTER MARY ELIZABETH

Sister Mary Elizabeth is the founder and driving force behind the AEGiS Web site, which she has made her life's work since the early 1990s. Before embarking on her religious vocation, she had a military career with the U.S. Navy. She served in Vietnam and worked over the years as a antisubmarine warfare technician, a support diver, a SCUBA instructor, and a electronics technician. She made her vows with the Sisters of Elizabeth of Hungary in 1988, and transferred to the Order of St. Michael in 1997.

HMI*: AEGiS is one of the best known HIV/AIDS resources on the Internet, and one of the oldest as well. Can you describe how AEGiS got its start?

Sister Mary Elizabeth: AEGiS actually had two beginnings. In 1986, Jamie Jemison, an Orange County resident, started the AEGiS (for AIDS Education General Information Service) bulletin board system [BBS]. Unfortunately, he was ahead of his time. Personal computers were just coming into vogue, and modems were extremely expensive. A 300 baud modem could cost as much as $1,200. Because of this, little interest was shown in the technology, and he closed the service down in early 1989.

In November 1989, I went to Missouri to herd cows. I met a couple of people with AIDS [PWAs] there, both of whom were geographically isolated from information and reliable health care, and I realized that an electronic bulletin board would be the perfect medium for disseminating information. I had been an aviation electronics instructor in the Navy, as well as an amateur radio operator, and I had, just prior to departing for Missouri, begun to put together an electronic bulletin board to publish information on transgender issues.

I returned from Missouri in March 1990, and begin putting together the HIV/AIDS Info BBS. Terry Travis introduced me to

*Interview conducted via e-mail, January 2001, by Liz Highleyman for HMI.

FidoNet, a kind of amateur or poor man's Internet, that had been created in 1986, by Tom Jennings. FidoNet, at its peak, connected some 66,000 electronic bulletin boards in over 60 countries.

The HIV/AIDS Info BBS was an instant success. Other BBS sysops (short for system operators) with an interest in HIV/AIDS quickly asked to share information with us, and a loose-knit network in approximately 40 countries formed over the next two years. Jamie heard about our work, and turned the AEGiS name over to us. Chris Quilter, one of our early board members, changed the name from AIDS Education General Information Service to AIDS Education Global Information System.

During the IX International AIDS conference held in 1993 in Berlin, the Global Electronic Network for AIDS (GENA) was formed, and members began identifying as GENA/aegis, GENA/hivnet, etc., to denote they were a network member. AEGiS became the cornerstone of the network.

HMI: How did AEGiS evolve from a BBS to its current format on the Web? How is it funded?

Sister Mary Elizabeth: In 1995, Bill Majors at Sandler Communications contacted me about porting our database of AIDS information to the newly created World Wide Web. The Internet, as you may recall, had previously been restricted to military and academic members. Bill put us in contact with Tom Sawyer at Roxane Laboratories [a large pharmaceutical company], and Tom agreed to provide funding. At the same time, Gale Dutcher and Anne White-Olson suggested that I apply for funding from the National Library of Medicine. Wynn and Rick Wagner volunteered to provide technical assistance, and AEGiS was ported to the Web in July 1996.

In 1999, Roxane provided additional funding, and I was able to hire Jeff Greer as our business manager. Jeff succeeded in obtaining funding from the John M. Lloyd Foundation, and he hired my first content assistant, Vanessa Robison, in April 2000.

Our current funding comes from Boehringer Ingelheim/Roxane, the National Library of Medicine, the John Lloyd Foundation, and our users. Our operational costs run about 20 percent or less of the costs of similar operations.

HMI: What are your thoughts about the shift of the Web away from free information toward commercialization? Will free information continue to play a role?

Sister Mary Elizabeth: According to most business reports, commercial medical information sites are having a very difficult time because people are not willing to pay for information over the Web. Their business models are now tending to focus within the medical profession itself, particularly in helping doctors and medical institutions integrate and help cut costs.

That opens room for the not-for-profit sites to continue what they have done all along: provide extensive medical information at no charge. The most important element here is that these sites build their academic and medical credibility.

HMI: What are your thoughts on the role played by the Internet in disseminating HIV/AIDS information? Does the Web, or the Internet more broadly, have unique aspects that make it better than other media for conveying this type of information?

Sister Mary Elizabeth: The "magic bullet" to cure or prevent HIV infection has not been found, and too many people who have or who are affected by HIV/AIDS are isolated by cultural, geographic, and economic barriers. In my opinion, our primary defense against HIV is the transformation of information into knowledge. To accomplish this, however, requires that information must be easily accessible and widely disseminated. The Internet, and FidoNet, to a lesser extent, have provided the means to accomplish this.

It was not until the advent of the World Wide Web that it became truly possible to make information easily accessible and widely disseminated in ways that often defy the imagination. Today, one can launch a Web browser, access a search engine, type in "AIDS," and retrieve hundreds of thousands of articles. Using a combination of key words, one can restrict a search to the point that precise articles can often be retrieved in seconds, from the privacy of one's home or office.

The beauty of this can be seen in a typical scientific journal or newsletter article from the AEGiS database. As you scroll through an article using your Web browser, you will see that all the references cited within the article are actually live hyperlinks to their source abstracts or full-text journal articles, making your research even more precise and timely. AEGiS has been linking references like this for a couple of years now, with the goal of having every article containing references linked in this fashion.

The Internet—but more so, the World Wide Web—has literally changed the way we communicate and perform literature searches. If the Gutenburg press was a small step for humankind, the Web was a giant leap, breaking down barriers to information access.

Chapter 13

Business, Labor Studies, and Economics

Alan Ellis

For those interested in conducting LGBTQ research in business, labor issues, management, organizational behavior, or economics, the following sites provide a solid starting point. Many of the Internet resources in the social sciences and law are also likely to be useful if you are conducting research on LGBTQ economic issues.

GENERAL SITES

Gay America Business Directory
http://gabd.com

The Gay America Business Directory lists businesses by state—click on the state of your choice on a map of the United States to access the listings for that state. The site requires businesses to pay a small fee to be listed, which appears to limit the number of businesses. However, if you are interested in researching queer-related businesses in a specific location or area, this site may provide you with a contact person or business that can guide you to other local resources.

Human Rights Campaign Worknet
http://www.hrc.org/worknet

The Human Rights Campaign's (HRC) Worknet provides a monthly update on workplace news and includes links to information about which corporations include sexual orientation in their nondiscrimination statements and which offer domestic partner benefits.

National Gay and Lesbian Task Force's Policy Institute
http://www.ngltf.org/pi

The National Gay and Lesbian Task Force's (NGLTF) Policy Institute's, described in Chapter 9, "Social and Biological Sciences," includes resources that focus specifically on employment issues.

Victory Magazine
http://www.webhelper.com/victorymagazine

Victory Magazine promotes itself as the premiere gay and lesbian business Web site. Here you will find an archive of articles on such topics as marketing and financial planning and empowerment. The site contains links to a directory of gay and lesbian businesses and a list of the top fifty fastest-growing gay and lesbian businesses, as well as other potentially interesting and valuable information. Parts of the site are still under construction.

LABOR STUDIES AND WORKPLACE ISSUES

Gay Workplace Issues
http://www.nyu.edu/pages/sls/gaywork

Sharon Silverstein, the co-author of *Straight Jobs, Gay Lives* (Touchstone Books, 1996), created this comprehensive listing of workplace-related resources that includes professional and business organizations by state and country, employee groups, and nonprofit and for-profit organizations that focus on workplace issues for the queer community. The site also includes a listing of publications on workplace issues (the list is current only as of October 1995, so you may want to go to one of the major online booksellers for a listing of more recent titles). There are also links to demographic information that you can use to conduct market research. Most of the links are current and quite useful, but the site and its content needs to be updated.

Lavender Collar
http://www.lavendercollar.com

Lavendar Collar provides workplace resources for the gay, lesbian, bisexual, and transgender communities. The site includes links to useful research sites with categories for employee groups, employers, the military, law enforcement, government, employment, organizations, workplace resources, legal issues, news, workplace issues, and transgender resources.

Pride at Work
http://www.prideatwork.org

Pride at Work is a constituency group of the American Federation of Labor and Congress of Industrial Organizations (AFL/CIO). Pride at Work seeks to create full equality for LGBTQ workers and to work with the organized labor movement to foster social and economic justice. The site includes tools for organizing and news about labor-related LGBTQ issues.

Queer Resources Directory: Workplace Listings
http://www.qrd.org/workplace

The Queer Resources Directory's (QRD) workplace listing is an excellent resource that includes numerous links to relevant online resources. The site includes links to employee groups and organizations specializing in workplace issues of concern to the queer community.

DOMESTIC PARTNERSHIP AND WORKPLACE NONDISCRIMINATION ISSUES

The best information regarding domestic partnerships continues to be available primarily in print and, if you are researching this topic, you may wish to obtain a copy of *Straight Talk About Gays in the Workplace* (The Haworth Press, 2000). A number of online resources can provide you with updated information about which organizations have nondiscrimination policies and which offer domestic partnership benefits.

Nolo Press
http://www.nolo.com/encyclopedia/articles/mlt/dp_benefits.html

Nolo Press offers a brief online article on domestic partner benefits, including a history of domestic partnerships and which private businesses, cities, and states were the first to offer such benefits. The article is a good introduction to the key issues concerning domestic partner benefits. Nolo Press has provided self-help legal resources for Americans since 1971, and is considered by many to be the most trustworthy source of such information.

Queer Resources Directory Sexual Orientation Nondiscrimination List
http://www.qrd.org/browse/sexual.orientation.nondiscrimination.list

Human Rights Campaign's WorkNet
http://www.hrc.org/worknet

These two sites provide listings of organizations that include sexual orientation as part of their nondiscrimination statements. Both are regularly updated and can be used to analyze the types of businesses and other organizations that have nondiscrimination policies. The WorkNet site includes a sample inclusive nondiscrimination statement. WorkNet also lists organizations that offer domestic partnership benefits and provides information on how to set up domestic partner benefits and nondiscrimination policies.

10 Percent
http://www.10percent.org/projectplan.html

The 10 Percent organization is "a community alliance of gay, lesbian, bisexual and transsexual citizens who are committed to effectuate the equitability of benefits and fairness in the law, for everyone."[1] This site provides information regarding strategies that one can use to encourage the adoption of domestic partnership benefits. The information is based on a number of sources, including a survey by KPMG Peat Marwick, a major consulting and accounting firm. You will also find a number of useful links on this site, including a link to a downloadable, 150-page report titled *The Domestic Partner Organizing Manual,* prepared by the NGLTF (the manual is also available from the NGLTF site at <http://www.ngltf.org>).

ECONOMICS

Institute for Gay and Lesbian Strategic Studies
http://www.iglss.org

The Institute for Gay and Lesbian Strategic Studies (IGLSS), mentioned earlier in the political science section of Chapter 9, "Social and

Biological Sciences" is a valuable resource for those conducting research on economic issues relevant to the LGBTQ communities. M.V. Lee Badgett, PhD, a noted economist, is currently the president of the board of IGLSS. IGLSS confronts a number of economic myths about the LGBTQ communities including the following:

> Myth: "Homosexual households had an average income of $55,400 . . . compared with a national average of $36,500. . . . This is not the profile of a group in need of special civil rights legislation It is the profile of an elite." Joseph Broadus, for the Family Research Council.
>
> IGLSS Fact: Our second publication destroyed this myth with scientific studies showing that gay people actually earn less than comparable heterosexuals.[2]

NON-WEB RESOURCES

Domestic Partner Benefits Mailing List
domestic@cs.cmu.edu

The purpose of this list is to facilitate networking about domestic partner benefits issues. To subscribe, send e-mail to <majordomo @cs.cmu.edu>.

Out4BizUSA
Out4BizUSA@gsb.stanford.edu

The purpose of this mailing list is to address nationally the job placement and school environment concerns of lesbian, gay, and bisexual masters of business administration (MBA) students. However, others are welcome to join as well, especially GLB MBA holders. To subscribe, send e-mail to <listproc@gsb.stanford.edu>.

NOTES

1. *10 Percent*. No creation date listed. 10 Percent. 8 December 2000. <http://www.10percent.org/our-mission.html>.
2. Institute for Gay and Lesbian Strategic Studies (IGLSS). No creation date listed. IGLSS. 19 January 2001. <http://www.iglss.org/support/about-iglss.html>.

Chapter 14

Community Resources

Melissa White

If you are conducting research on community issues and resources, you will find many leads at the two major portals, Gay.com (http://www.gay.com) and PlanetOut (http://www.planetout.com), as well as at the Web sites of the two major LGBT political organizations, the Human Rights Campaign (http://www.hrc.org) and the National Gay and Lesbian Task Force (http://www.ngltf.org). Following are sites of several organizations that address a variety of needs and community-based projects created by and for LGBTQ people.

LGBTQ COMMUNITY-BASED RESOURCES

Children of Lesbians and Gays Everywhere
http://www.colage.org

As the Web site of Children of Lesbians and Gays Everywhere (COLAGE) states, "COLAGE is the best (well okay, the *only*) support and advocacy organization for daughters and sons of lesbian, gay, bisexual, and transgender parents." This excellent resource offers an archive of COLAGE's newsletters, an interactive story for young people, information about publications and videos of interest to families with one or more queer parents, a site-search function, e-mail lists, contact information for COLAGE chapters in sixteen U.S. states and two foreign countries, and much more.

Funders for Lesbian and Gay Issues
http://www.lgbtfunders.org

Formerly known as the Working Group on Funding Lesbian and Gay Issues, this organization seeks to increase the amount and variety of resources available for LGBTQ projects and organizations and to educate grantmakers and philanthropists about LGBTQ issues. This site provides online access to a very useful publication entitled *Funders of Lesbian, Gay, Bisexual and Transgender Programs: A Directory for Grantseekers,* which includes links to many of these funders' web sites. Another unique publication downloadable from the site is *Creating Communities: Giving and Volunteering by Gay, Lesbian, Bisexual and Transgender People.* In partnership with the Women's Funding Network (http://www.wfnet.org), Funders for Lesbian and Gay Issues is preparing a new publication that will include unique research about lesbian donors' motivations for giving and ways to effectively make appeals to these donors. Additional general resources for grantseekers (such as the more extensive collections of the Foundation Center at <http://www.fdncenter.org>) and basic tips on grantseeking are also provided.

Gay-Straight Alliance Network
http://www.gsanetwork.org

Founded in 1998 and led by youth, the Gay-Straight Alliance Network (GSA Network) connects and empowers San Francisco Bay Area youth activists fighting homophobia by organizing themselves in student-run gay-straight clubs found in many high schools. The GSA Network offers training, leadership development, and other kinds of support to strengthen existing GSAs and to help young people start new ones. At the site, you can learn how to start a GSA in your community, read the GSA Network's archive of newsletters, and find out about leadership training opportunities for youth. At the time of this writing, a project was underway to create a national network of Gay-Straight Alliances.

Gay, Lesbian and Straight Education Network
http://www.glsen.org

The Gay, Lesbian and Straight Education Network (GLSEN), described in Chapter 10, "Arts and Education," works to find effective ways to resolve problems that youth all too often face in schools because of their real or perceived sexual orientation or gender identity. GLSEN's site offers information about local chapters, current campaigns, ways to become an activist to help make schools safe for LGBTQ youth, a bookstore, events calendar, and news.

Harvey Milk Institute
http://www.harveymilk.org

The Harvey Milk Institute (HMI) is a community-based organization that conducts a variety of educational, art, and cultural activities in the San Francisco Bay Area and beyond. On the Web site, you can find out about issues pertinent to contemporary queer communities and peruse the latest catalog of over one hundred courses, workshops, and events created by and for LGBTQ people.

International Association of Lesbian, Gay, Bisexual, Transgendered Pride Coordinators
http://www.interpride.org

The International Association of Lesbian, Gay, Bisexual, Transgendered Pride Coordinators, Inc. "exists to promote lesbian, gay, bisexual, and transgender pride on an international level, to increase networking and communication among Pride groups, to encourage diverse communities to hold Pride Events, and to act as a source of education." Visit this site for information about pride events throughout the world.

National Association of Lesbian and Gay Community Centers
http://www.gaycenter.org/natctr

Since 1994, the National Association of Lesbian and Gay Community Centers has been linking queer community centers nationwide and promoting informed voting among LGBTQ populations. The National Directory of Lesbian, Gay, Bisexual, and Transgender Community Centers is one project of this group. Visit the site to view the directory or to apply for membership for your center and have it added to the directory.

National Deaf Queer Resource Center
http://www.deafqueer.org

The National Deaf Queer Resource Center is a nonprofit resource and information center founded and directed by deaf queer activist Dragonsani Renteria. The Web site, launched in September 1995, offers the most comprehensive and accurate information about the deaf lesbian, gay, bisexual, and transgendered community.

Parents, Families, and Friends of Lesbians and Gays
http://www.pflag.org

Parents, Families, and Friends of Lesbians and Gays (PFLAG) provides support, education and advocacy for LGBTQ individuals and their families. Using this site, you can find a chapter close to your community—or learn how to start one—order books that help families work constructively through coming-out issues, read PFLAG's policy statements, sign up for an e-mail alert list, or become a member.

NON-WEB RESOURCES

Hotlines

- GLBT Hate Crimes Hotline
  (800) 616-HATE
- Gay and Lesbian National Hotline
  (888) THE-GLNH
- LYRIC Youth Talk Line
  (800) 246-PRIDE

A peer talk line for LGBTQ youth twenty-three and under. Available Monday-Saturday, 6:30-9:00 p.m.

INTERVIEW: KEVIN SCHAUB

Kevin Schaub is a co-editor of this guide; his biography is listed in the About the Editors section.

HMI*: When did you start with the Harvey Milk Institute?

Kevin Schaub: In 1994, my then-partner Jonathan Katz, a cofounder of the Harvey Milk Institute (HMI), and I hosted a long series of community meetings in our home about what HMI could be. I had no intention of taking a long-term leadership role at the time. I was volunteering once a week in the office and teaching French for the Queer Traveler. We opened our doors in January 1995. Chaos ensued, and it was obvious that someone needed to take charge of the organization

*Interview conducted on December 14, 2000, by Alan Ellis for HMI.

administratively. I cheerfully volunteered my services as administrative director thanks to having a flexible "real" job.

HMI: When did you become the official Executive Director?

Kevin Schaub: A few months later. There I was.

HMI: At the time did you use the Internet much?

Kevin Schaub: No. It wasn't until 1997 that we got a computer with a modem.

HMI: How has the Web affected your work and that of HMI?

Kevin Schaub: Once we got online, we moved rapidly and had a limited Web site up by the end of 1997. In 1997, we had someone volunteer to build us a Web site that could display our courses online for our Spring 1998 catalog. We've since upgraded the site, and we now offer our students easy online registration and the capability to make donations online. Now, over half our registrations come through our Web site. It has enormously improved our ability to serve our students.

HMI: Are there other ways that you use the Internet to facilitate your work?

Kevin Schaub: We use it to research artists, speakers, and others who might offer events or courses, and to research other organizations with whom we might collaborate. In addition, people contact us for information and referrals. Before we had Web access, I had a more difficult time accessing information relating to queer studies and culture. The Web has greatly increased our ability to share information and respond to our constituency.

HMI: Are there specific Web sites that you use a lot?

Kevin Schaub: Well, I go to <http://www.gogos.com> a lot but that's not really about HMI. Actually, we use Queer Arts Resource (http://www.queer-arts.org), and the Queer Cultural Center (http://www.queerculturalcenter.org), the Gay, Lesbian, Bisexual, and Transgender Historical Society of Northern California (http://www.glbthistory.org), and the National Gay and Lesbian Task Force (http://www.ngltf.org). I also use Poets and Writers, Inc. (http:// www.pw.org) to get biographies of poets and writers around the globe, some of whom we invite to read at festivals.

More and more authors and artists have their own Web sites as well, for example, Jewelle Gomez (http://www.jewellegomez.com), which helps me to contact them and organize events and help with publicity.

We also use Google a lot. This search engine is especially helpful because I am often looking for specific quirky subjects and it helps to put things into context. If there is a good set of links on a site, that can be an amazing resource for me.

HMI: My understanding is that you get a lot of requests for information via the Internet. What types of questions do people ask HMI?

Kevin Schaub: The number one request goes something like this, "I'm doing a paper on Harvey Milk, what are good sources of information?" Generally, they have seen *The Times of Harvey Milk* and have decided to do a paper on him, but don't know where to find information about gay men at their high school or college library.

HMI: What do you tell them?

Kevin Schaub: Usually, they're in quite a rush to complete their paper. I refer them to the GLBT Historical Society of Northern California site and to Randy Shilt's book, *The Mayor of Castro Street*. Another great resource on the Web for getting a more personal look at Harvey Milk as a man, not just as an icon, is Uncle Donald's Castro Street's at <http://www.backdoor.com/CASTRO/milkpage.html>. I also recommend that they do an online search using Google.

HMI: What other questions do people ask?

Kevin Schaub: Generally anything related to queer research that they couldn't find any resources on in the library. In reality, however, it's often the case that people don't know how to do research, or the book is checked out or otherwise unavailable. What's surprising is that many people don't think of doing a simple online search or don't know what keywords to use. Moreover, if they do know how to do research, they can't determine if what they find is a credible source, so they sometimes call to ask us for references and other resources that we consider to be legitimate.

HMI: Do you have any other thoughts about the Internet and its role in conducting queer research?

Kevin Schaub: It's mind-boggling to realize how much information is out there. The key is to figure out how to access it and how to filter out the good stuff from the fluff. I think about the amount of information I absorb daily now because of the Internet as compared to four years ago, and how much more easily I can find specifically what I want without being exposed to the "sports" section of the daily paper. For example, I subscribe to a number of free news sources (e.g., *New York Times*) and Web alerts from such organizations as NGLTF and GLAAD. What's great about these sites is that even though I live in San Francisco, where there are four free LGBTQ newsweeklies, I could move to North Dakota and still be connected to the queer universe.

Chapter 15

Media and News

Mark Menke

For many research topics, an understanding of current issues and trends in the LGBTQ communities is essential. In addition, a review of the contents of one or more of the following sites can help define and refine a research topic.

MEDIA RESOURCES

The Advocate Online
http://www.advocate.com

The online version of *The Advocate* magazine contains some content from the current issue on newsstands. It also has updated news briefs and links to columnists like Michelangelo Signorile. The site includes archives of past stories and news arranged by topic rather than by date (e.g., Boy Scouts, Election 2000, transgender, etc.).

Gaywire
http://www.gaywire.net

Gaywire, a project of Gay.com, features current and archived releases on a wide range of news topics. Users can post their own releases. Gaywire features an affiliates list of other sites that embed Gaywire within themselves. This can function as a small portal to start queer media research on a more localized level.

Gay and Lesbian Alliance Against Defamation
http://www.glaad.org

The Gay and Lesbian Alliance Against Defamation (GLAAD) was "formed in New York in 1985 and began by protesting *The New York*

Post's blatantly offensive and sensationalized stories about AIDS. Its mission was to improve the public's attitudes toward homosexuality and to put an end to violence and discrimination against lesbians and gay men." GLAAD's Web site is an excellent resource for articles, essays, news updates, and publications, and includes the organization's weekly *GLAAD Lines,* which offers news, tips, and information to media professionals, and the biweekly *GLAAD Alerts,* which notifies the LGBTQ communities of media defamation and calls to action.

Gay and Lesbian Radio Network
http://www.gaybc.com

This is the site of the "world's gay and lesbian radio network." Check out the site for a list of shows covering a wide range of topics. GayBC radio is broadcast from Seattle, Washington, and includes shows hosted by Grethe Cammermeyer and other well-known LGBTQ personalities.

Michelangelo Signorile
http://www.signorile.com

This is the official home page for New York-based journalist Michelangelo Signorile, a columnist for *The Advocate* and the author of several books including *Queer in America* and *Outing Yourself.* It contains archives of past articles and discussion threads.

PlanetOut.com
http://www.planetout.com

PlanetOut—a major gay portal site—features daily news updates and ongoing features broken up by topic areas. Many news items link to *The Advocate's* Web site. Visitors can sign up for a free membership to receive e-mail digests of daily headlines, weekly entertainment news, etc.

Queer Resources Directory
http://www.qrd.org/qrd/media

Check out the media section of the Queer Resource Directory, which offers a very useful and extensive list of links. (QRD is described in Chapter 3, "Major LGBTQ Internet Research Tools.")

Rex Wockner News
http://www.wockner-news.com

This site features journalist Rex Wockner's archive of his famous "quote/unquotes." Wockner's news service has been used by many local and national LGBTQ publications since 1986. You can search his site to find useful quotes and national and international news articles. To view some of his archived articles, visit: <http://www.qrd.org/qrd/www/world/wockner.html>. To see Wockner's weekly (Friday) updates on politics and culture, visit <http://www.planetout.com/wocknerwire>.

NON-WEB RESOURCES

GLB-NEWS
glb-news@brownvm.brown.edu

GLB-NEWS is a repository for news of interest to LGBTQ individuals. It offers readers information about news and events, action alerts, conference information, and more. This edited list is not intended to facilitate conversation between readers. When an article or message is submitted, it is reviewed by a moderator to determine whether it will be distributed to the list. To subscribe, send e-mail to <listserv@brownvm.brown.edu>.

Press Pass Q
http://www.qsyndicate.com/PressPassQ.htm

Press Pass Q (PPQ) is an online "newsletter for the gay and lesbian press professional." It features original articles about issues and trends concerning the queer press, as well as updates of changes at LGBTQ publications. PPQ and the Vice Versa gay press awards were recently acquired by Rivendell Marketing.

Usenet Newsgroups

For news from a "homosexual viewpoint," go to **alt.journalism.gay-press**.

Chapter 16

The Queer Internet: One Student's Experience

David Brightman

Shocking as it will seem to many of this book's readers, I did not encounter the Web until I was a student in college. The University of California at Berkeley gives every student the privilege of an e-mail account and access to the Web in the computer lab, as I suppose most others schools do. I was not much impressed with the Web at first, but that is because I had no use for it. It just seemed like a bunch of pointless graphics, with very little to offer the individual who clicked through to them. Then I signed up for a course on LGBTQ culture and health in the School of Public Health, and one of the first assignments in the course was to do some research on the Web. It was a very simple assignment, but a good one. We were to look for information on a topic, and make notes on how many links it took to get there, and the path that we took to find it. I no longer remember what I looked for, but I remember being quite intrigued by what I clicked through on my way to useful information, and trying to imagine what other people experienced as they clicked their way toward what they needed to know.

In recent years, there has been a lot of talk about how the Internet will eventually transform everything about the way we live. Although the Internet has not changed everything about the way I live, it has become an essential research tool. But I still have to decide whether what the Internet shows me is helpful or a waste of time. As David Rothenberg opined on the randomness of the Web, "Chance holds sway, and more often misses than hits."[1] The Internet holds enormous amounts of information for queers, but that information may be of little use to us without some frame of reference to give it meaning. In their book, *The Social Life of Information,* authors John Seely Brown and Paul Duguid maintain that information is of little value without this social context, pointing out that the "social forces, always at work, within which and against which individuals configure their identity . . . create not only grounds for reception, but grounds for interpretation, judgment, and understanding."[2] Attention to the social forces that shape our lives—including what we encounter on the Internet—is really a fundamental prerequisite to any research work that queers undertake, but this frame of reference becomes especially important when evaluating the disembodied information that we find online.

My own frame of reference seemed to diminish the importance of Web-based information, at least when I first began to use it. Internet research was not especially important to me as I pursued my interdisciplinary studies of sexuality and gender at UC Berkeley, because I had access to most of what I needed to know in plain old printed books and journals. Some print sources stood out as being particularly important, such as the fabulous files in the archives of the Gay, Lesbian, Bisexual, and Transgender Historical Society of Northern California. I went there to research the history of the transsexual/transgender community in San Francisco for a term paper, and the archivist gave me several boxes of fascinating material to look through (probably from the papers of Louis Sullivan, a founding member of the Society and pioneer trans-historian), including wonderful early magazines and newsletters from the 1960s (which reminded me somewhat of the 'zines produced by queer cultural activists in the 1990s, in terms of close connection to the community and their challenge to basic assumptions of mainstream society). This access to the collections could conceivably be approximated online someday, and I would applaud that development. Of course, the GLBT Historical Society does have a wonderful Web site (http://www.glbhistory.org), but they do not have every item in the collection available in digital form. Furthermore, there is a qualitative difference in the experience that would be lost without the physical presence of collection and archivist. I once bumped into Susan Stryker, the society's Executive Director, at a Gay Freedom Day exhibit of items from the collection. We chatted about the display, and I learned fascinating historical information (okay, gossip, but it was important historical gossip about an important early philanthropist in the trans community) that might not have come up in an online class or interview. In a way, this encounter enhanced my appreciation of the Web site, because I knew what the organization had to offer, but at the same time it illustrated the limits of what can be made available to anyone online.

Since graduation, I have gained a greater appreciation of the Internet as a research tool. I am immensely grateful to have so much information at my fingertips, and appreciate the ease of e-commerce and e-mail. Still, it is a tool I most often use to locate some resource that exists in physical as well as virtual space: an organization, institution, book, or person. In most cases, the resource's identity is not formed primarily online, however important the ability to find it there may be. I worked for a time with a company that develops courseware for Web-based training, and some of the work we did was informed by Internet research. But to a large extent, we still turned to books for various kinds of information and assistance in our work, and I do not believe that this company is at all unusual in this respect. The founders of the company were highly respected in their field, and as a result were often asked to speak at conferences or write books. If you believe the futurologists, by now we ought to have abandoned tedious, inefficient bodily travel in favor of virtual travel, and replaced clunky old books with "just-in-time" online text delivery. Neither has been abandoned, nor have I seen any convincing sign that we are likely to do so any time soon.

What will happen is that the Web will become a more and more useful tool. As I was writing this, the latest issue of an e-mail newsletter I subscribe to arrived, with news of new specialized search agents that will bring more relevant information in response to user queries. It is not the abandonment of the context in which we really live, but rather a deeper responsiveness to that context, which will make Internet research an increasingly important part of our lives.

NOTES

1. Rothenberg, David (1997). "How the Web Destroys the Quality of Students' Research Papers." *Chronicle of Higher Education,* August 15, p. A44

2. Brown, John Seely and Duguid, Paul (2000). *The Social Life of Information.* Boston, MA: Harvard Business School Press, p. 139.

Index

A Credit to Her Country, 84
ABGender, 69
About.com, 25
Above and Beyond, 61
Access Denied, 47
Adelaide University, 102
Adobe Acrobat, 70
Adrenal hyperplasia, 137
Adultcheck, 25
The Advocate
description of, 161
references to, 34, 35, 36, 40, 89, 112, 123, 162
Affirmation, 91
AIDS. *See* HIV/AIDS
AIDS Education Global Information System (AEGiS)
description of, 138
interview with founder, 143-146
references to, 2
AIDSLINE, 142
The Alan Turing Project, 46
Al-Fatiha Foundation, 95, 96
Alliance for Computers and Writing (ACW), 20, 21
al-Qur'an, 99
Altavista, 6, 7, 8, 50
alt.gossip.celebrities, 14
alt.journalism.gay-press, 163
alt.polyamory, 81
alt.sex.bondage, 80
Altsex.org, 72
Ambiguous genitals, 137
America Online (AOL), 15, 31, 34, 36, 42, 66
American Academy of Pediatrics, 141
American Anthropological Association, 102
American Cancer Society, 141
American Educational Gender Information Service (AEGIS), 62
American Family Association, 98
American Federation of Labor and Congress of Industrial Organizations (AFL-CIO), 149
American Heart Association, 141
American Medical Association, 141
American Political Science Association (APSA), 103
American Psychoanalytic Foundation, 105, 106
American Psychological Association
description of, 106
Policy Statements on Lesbian and Gay Issues, 107
American Psychological Society, 106
American Sociological Association, 108
Androgen sensitivity, 137
Angles, 51
Annual Review of Gender and Sexuality Law, 127
Annual Review of Sex Research, 74
Anthropological Resources on Sexuality Issues, 102
Anthropology, 101, 102, 103
Anything That Moves, 55
Archie, 16
ARPAnet, 5, 45
Art in America, 123
Art.com, 125
Arts, 115
Ask Jeeves, 8
Austin, Kimberly, 121

Bacon, Francis, 84, 86
Badgett, M. V. Lee, 151
Baldwin, James, 86
Barnett, David, 14, 52
Bawer, Bruce, 26
Bay Area Reporter, 31, 66
Bay Area Sex Worker Advocacy Network, 79

BDSM
 references to, 10, 29, 49, 71
 resources, 72-77
Bean, Joseph, 75
Bennett, Jeff, 31
Berenstein, Rhona
 interview with, 30-33
 references to, 24
Berkeley Hass School of Business, 30
Berners-Lee, Tim, 6
Bgay, 28
Bi the Way, 56
Bibliography on Human Sexuality, 73
BiCafe, 56
BiFem Net, 58
Bint el Nas, 87
Biological sciences, 101, 109, 110
Bi.org, 56
Bisexual Action on Sexual Health, 136
Bisexual health issues, 136
Bisexual Resource Center
 description of, 57
 Health section, 136
Bisexual Resource Guide, 57
Bisexual Youth, 58
Bisexual.org, 56
Bisexuals
 references to, 49, 55
 resources, 55-59
The Black List, 87
BLK Homie Pages, 87, 113
The Body, 138, 139
The Body Electric, 90, 91
Boehringer Ingelheim/Roxane, 144
Boston Bisexual Women's Network, 58
Boston University, 30
Brazilian travestis, 49
Breast Cancer Action, 141
British University, 73
Brown, John Seely, 165
Brown University
 Bi Lists, 58
 references to, 65
Buckmire, Ron, 41
Buddhism, 89, 91
Bulletin Board Systems (BBSs), 11
Bulletin of Experimental Treatments for AIDS (BETA), 140
Business, 147
Caja, Jerome, 116
Cameron, Dan, 123
Cameron, Paul, 107
Canadian Lesbian and Gay Archives, 83
CANCERLIT, 142
Carol Wilson Award, 39
Case, Steve, 42
Cather, Willa, 86
Catholic Church, 64, 92
Catholic Lesbians, 91
Catholic Theology Library, 92
CellNet Data Systems, 30
Center for Lesbian and Gay Studies (CLAGS), City University of New York (CUNY)
 description of, 50, 103
 Gender and Sexuality Studies Mailing List, 54
 references to, 19, 112
Centers for Disease Control and Prevention, 141
Chat forums, 15
Chicanos/as, works on gay, lesbian, bisexual, and transgendered, 88
Children of Lesbians and Gays Everywhere (COLAGE), 153
Christianity, 89, 91-95
Citations
 American Psychological Association, 18
  Copyrights and Permissions, 21
  Electronic Reference Formats Recommended by the American Psychological Association, 21
 Basic CGOS Style, 21
 Beyond the MLA Handbook: Documenting Sources on the Internet, 21
 The Columbia Guide to Online Style, 21
 Electronic Style: A Guide to Citing Electronic Information, 22
 Electronic Styles: A Handbook to Citing Electronic Information, 22

Citations *(continued)*
  International Federation of Library Associations and Institutions, *Library and Information Science: Citation Guides for Electronic Documents,* 21
  Modern Language Association (MLA), 18
    MLA Handbook for Writers of Research Papers, 19
    MLA-Style Citations of Electronic Sources, 20
    Wired Style: Principles of English Usage in the Digital Age, 22
City TV, 36
Clitoromegaly, 137
Coalition for Positive Sexuality, 78
Cochran, Susan, *xiii*
Columbia University, 78
Commercial Sex Information Service, 79
Community resources, 153
CompuServe, 36
Cornell University, 85
Corrine, Tee, 116
Crane, Nancy, 22
Creating Communities: Giving and Volunteering by Gay, Lesbian, Bisexual and Transgender People, 154
Credibility, 16, 17
Crew, Louie, 54

Dahir, Mubarak, 25
Dartmouth College, 44
De Cecco, J., 109
DECPLUS, 46
DeGeneres, Betty, 24, 38
Diamant, Allison, *xiii*
Diaz, Jean "Ambar," 30
Dickinson, Emily, 86
Different Loving, 75
Digital defamation, 46
Digital divide, 46
Digital Queers, 40, 41
Dignity USA, 92
Direct Hit, 6, 8
Division 44 Annual Society for the Scientific Study of Sexuality (SSSS), 74
DMOZ, 10
Dogpile, 7, 8
Doherty, Will
  interview with, 44-48
  references to, 29
Domestic Partner Benefits mailing list, 151
The Domestic Partner Organizing Manual (NGLTF), 150
Domestic Partnership issues, 149, 150
Drag kings, 49
Drkoop.com, 141
Dubois, Cleo, 75
Duguid, Paul, 166
Duke University, 72, 115
Dutcher, Gale, 144

E! Online, 11
Early genital surgery, 137
Economics, 147, 150, 151
Education, 117
Educational Opportunities in Human Sexuality, 74
Elder Lesbians, 84
Elderkin, Mark
  interview with, 30-33
  references to, 24
Electronic Frontier Foundation, 44
  Extended Guide to the Internet, 5
Electronic Journal of Human Sexuality, 72
Electronic Mailing Lists
  description of, 12
  references to, 18
Elia, J., 109
Elizabeth, Sister Mary
  interview with, 143-146
  references to, 138
Ellis, A., 109, 110
Episcopal Church, 92
Epstein, Rob, *xvii*
Erratic Impact, 129
Ethnic studies, 87-88
The Eulenspiegel Society, 75
Excite, 8, 31

Federal Election Commission, 111
Fenway Community Health Center, 136
FidoNet, 144, 145
Fierstein, Harvey, *xvii*
File Transfer Protocol (FTP), 15, 18
Fine Arts Museum of San Francisco Web site, 123
Flash animation, 70
Flemming, Tom, 132
Fordham University, 94
The Foundation Center, 154
Frameline, 11
Free Agent for Windows, 14
FTM Information Network, 61, 69
FTM International, 62
Funders for Lesbian and Gay Issues, 153
Funders of Lesbian, Gay, Bisexual, and Transgender Programs: A Directory for Grantseekers, 154

Garnets, Linda, *xiii*
Gay, Lesbian, and Bisexual People of Color, 98
Gay, Lesbian, and Straight Education Network (GLSEN), 115, 117, 154
Gay, Lesbian, Bisexual, and Transgender Historical Society of Northern California
  description of, 83, 84
  references to, 36, 157, 158, 166
Gay America Business Directory, 147
Gay and Lesbian Advocates and Defenders, 127
Gay and Lesbian Alliance Against Defamation (GLAAD)
  description of, 161, 162
  Digital Media Resource Center in San Francisco, 44
  GLAAD Alerts, 162
  GLAAD Lines, 162
  Internet Leadership Award, 34
  Media Award, 67
  references to, 30, 31, 41, 128, 159
Gay and Lesbian Arabic Society, 87
Gay and Lesbian Association of Choruses (GALA), 48
Gay and Lesbian Medical Association, 131
Gay and Lesbian National Hotline, 156
Gay and Lesbian Radio Network, 162
The Gay and Lesbian Review Worldwide, 51
Gay and Lesbian Victory Fund, 105, 111
Gay Buddhist Fellowship, 91
Gay Christians, 98
Gay Crawler, 28
Gay History and Literature, 84
Gay Men's Health, 133
Gay Men's health concerns, 134
Gay Men's Health Crisis, 139
Gay Men's Health Summit, 133
Gay Muslims, 98
Gay Pagans, Gay Witches...Gay Witchcraft, 97
Gay Science: The Ethics of Sexual Orientation Research, 109
Gay Workplace Issues, 148
Gay.com
  description of, 24
  Health Channel, 131
  interview with co-founder, 30-33
  references to, 2, 15, 23, 25
GayHealth, 132
Gayhistory.com, 84
Gay/Lesbian Politics and Law, 103
Gay/Lesbian/Queer Social Science List, 110
GayNet, 45
Gays and Lesbians in the Democratic Process, 110
Gayscape, 28
Gay-Straight Alliance Network, 154
GayWellness, 134
Gaywire, 161
GayWired
  bibliography, 108
  description of, 25, 26
Gender Education and Advocacy
  description of, 62
  Gender Advocacy Internet News (GAIN), 62
  interview with Webmistress, 66-70
GenderLine, 69

Gender.org, 2, 69
GenderPAC, 62, 63
General consumer health sites, 140-141
General Magic, 34
Georgetown Journal of Gender and Law, 127
Gerber Hart Library, 84, 85
Getty Institute Web site, 123
Gibaldi, Joseph, 19
Gill Foundation, 125
Gillian Anderson Estrogen Brigade, 28-29
GLB-News, 163
GLBT Hate Crimes Hotline, 156
GLBT Health Access Project, 132
Global Electronic Network for AIDS, 144
GLQ: A Journal of Lesbian and Gay Studies, 72
Go Ask Alice, 78
Go Network, 9
Godhatesfags.com, 97
Golebiowska, Ewa, 111
Gomez, Jewelle, 158
Google, 6, 9, 11, 14, 37, 111, 141, 158
Gopher, 11, 16, 18
Grady, Jim, 122
Green, Bert, 122
Greer, Jeff, 144
Gutenberg press, 146

Hadith, 99
Hale, Constance, 22
Hall, Lesley, 20
Halsall, Paul, 94
Harnack, Andrew, 21
Harrison, Barry
  interview with, 117-125
  references to, 116
Harry Benjamin International Gender Dysphoria Association, 137
Harvey Milk Institute
  description of, *xvii*, 155
  interview with Executive Director, 156-159
Health, 131
Health Central, 141
Henkin, William, 75
Herek, Greg, 101, 102, 109
Hermaphrodites with Attitude, 64
HIV Insite, 139
HIV/AIDS
  references to, 32, 40
  resources, 73, 74, 101, 138-140
HIV/AIDS Treatment Information Service, 139
HIVandHepatitis.com, 139
Homo Promo, 35
Homorama, 25, 26
HOOK Online, 79
Horizons Foundation, 87, 125
Hotel Tax fund, 121
Hotlines, 156
Human Relations Area Files (HRAF), 73
Human Rights Campaign
  references to, 127
  Worknet, 147
Human Sexuality Collection, 85
Hypospadias, 137

ICQ, 15
If You Seduce a Straight Person, Can You Make Them Gay?, 109
Iman, 98
Independent Gay Forum, 26
Indiana University, 103
Infoseek, 9
Ingenta, 28
Ingersoll Gender Center, 137
Institute for Gay and Lesbian Strategic Studies, 51, 104, 111, 150
Institute for the Advanced Study of Human Sexuality, 72
Integrity USA, 92
International Association of Lesbian, Gay, Bisexual, Transgendered Pride Coordinators, 155
International Federation of Library Associations and Institutions. *See* Citations
International Foundation for Gender Education (IFGE), 63
International Gay and Lesbian Human Rights Commission, 104
International Gay and Lesbian Review, 86

The International Journal of Transgenderism, 19, 63
International Lesbian and Gay Association, 104, 112
Internet Explorer browser, 14
Internet Relay Chat (IRC), 11, 14
Intersex Society of North America (ISNA), 64, 137
Intersexuals
  references to, 49
  resources, 61-66
It's Time, America, 64

Jacob Hale's Rules, 64
The James C. Hormel Gay and Lesbian Center, 51
Jarman, Derek, 116
Javier, Loren, 47
Jemison, Jamie, 143
Jennings, Tom, 144
John M. Lloyd Foundation, 144
Joint Task Force on Professional Practice Guidelines, 107
Journal of the American Medical Association, 109, 142
Journal of the Gay and Lesbian Medical Association, 132
Journal of Sex Research, 74
JRI Health, 137
Judaism, 89, 95

Katz, Jonathan, 156
The Kinsey Institute, 72
Klein, Fritz, 57
Klein Sexual Orientation Grid, 57
Kleppinger, Gene, 21
Klinefelter Syndrome, 137
Klorese, Roger
  interview with, 44-48
  references to, 29
KPMG Peat Marwick, 150
Kramer, Joseph, 90

La Voz de Esperanza, 88
Labor studies, 147, 148, 149
Lambda Legal Defense and Education Fund
  description of, 128
  references to, 26, 127
The Lancet, 142
Latinos/as, works on gay, lesbian, bisexual, and transgendered, 88
Lavender Collar, 148
The Lavender Web: LGBTQ Resources on the Internet, 52
Law, 127
Learn the Net, 16
Leather Archives and Museum, 75
The Leather Page, 76
LeatherQuest, 76
LeatherWeb, 76
Leonsis, Ted, 42
Lesbian and Gay Caucus for Political Science, 105, 112
Lesbian Avengers, 28
Lesbian Health and Homophobia, 134
  Lesbian Health and Homophobia: Perspectives for the Treating Obstetrician/Gynecologist, 134
The Lesbian Health Fund of the Gay and Lesbian Medical Association, *xiv*
Lesbian Health Links, 135
The Lesbian Herstory Archives, 85
The Lesbian History Project, 85
Lesbian Mailing Lists, 28
Lesbian Mothers Support Society, 135
Lesbian.com
  description of, 26
  Health and Wellness, 134
  references to, 25
Lesbian.org
  description of, 27
  Introduction to the Internet, 16
LesbiaNation, 25
Lesbians in the Visual Arts, 115
Lesbians of Color, 88
LesbianWellness, 135
Leslie Lohman Art Foundation, 115
LGBT Schools and Libraries Project, 46
LGBT SWAT Team, 46
LGBT+ Internet Mailing Lists, 29
Li, Xia, 22

Listserv, subscription information, 12
LLEGÓ, 88
Log Cabin Republicans, 105
LookSmart, 9
Los Angeles Gay and Lesbian Center, 31
Lourde, Audre, 86
Lutherans Concerned, 92
Lycos, 10, 50
LYRIC Youth Talk Line, 156

Magic Cap 1.0, 34
Majordomo, subscription information, 12
Majors, Bill, 144
Marcus, Eric, 26
Massachusetts Institute of Technology (MIT), 29
  Gays and MIT (GAMIT), 45
  Media Lab, 34, 37, 39
The Mautner Project for Lesbians with Cancer, 135
Mayo Clinic, 141
The Mayor of Castro Street, 158
Mays, Vickie, *xiii*
McCarthy Era, 84
McKim, Marianna, 53
McMaster University Health Sciences Library, 132
Medicine, 131
Medina, Dennis, 88
MEDLINE, 142
Methodist, 91
Metropolitan Community Church (MCC), 93
Micropenis, 137
Milk, Harvey, *xvii,* 158
Miller, Penny, 111
The Mining Company, 25
Minority Health Research Catalog, 73
misc.fitness, 143
misc.health, 143
Mitchell, R., 109
Modern Language Association (MLA). *See* Citations
Monash University, 109
Moran, Michelle, 42
More Light Presbyterians, 93
Mormon, 91
Moscone, George, *xvii*
Mozilla, 10
Mujadarra Girls, 87
Murphy, T., 109
Murray, Stephen O., 26
Musafar, Fakir, 75
Muslim, 89, 95, 96
Muslim Gay Men, 99
My Dear Boy: Gay Love Letters Through the Centuries, 84
The Myth of the Modern Homosexual: Queer History and the Search for Cultural Unity, 84

Nasalam, 90
National Association for Research and Therapy of Homosexuality, 107
National Association of Lesbian and Gay Community Centers, 155
National Cancer Institute, 141
National Center for Lesbian Rights (NCLR), 128
National Coalition for Sexual Freedom, 73
National Deaf Queer Resource Center, 155
National Gay and Lesbian Task Force (NGLTF)
  Policy Institute, 102, 148
  references to, 41, 111, 157, 159
National Institute of Allergy and Infectious Diseases, 141
National Institute for Environmental Health, 141
National Institutes of Health, *xiv,* 108, 141
National Library of Medicine, 144
National Organization of Women (NOW)
  description of, 105
  SM Policy Reform Project, 20
National Science Foundation, NSFnet, 5
National Semiconductor, 30
National Transgender Advocacy Coalition, 64, 65
Net Queery, 42
Netscape browser, 14, 31

Network Equipment Technologies, 30
New England Journal of Medicine, 142
New England School of Law Gay, Lesbian, Bisexual and Transgender Caucus, 129
New Museum of Contemporary Art, 123
The New School, 36
New York Public Library, 52
New York Times Online, 123, 159
Newman, Felice, 27
Newswatcher for Mac, 14
NewsXpress, 14
NextMonet.com, 125
Nicolosi, Joseph, 107
Nolo Press, 149
Nomenus, 96
North Beach, 84
Northern Light, 10, 118
Norton, Rictor, 84

Occidental College (Los Angeles), 41
O'Hanlan, Kate, 134
Olsen, Jenni, 35
The One Institute, 86
O'Neal, Sean, 68
Online Partners, 30
Online Policy Group
  interview with Executive Director, 44-48
  references to, 29
Online Service Provider Assessment Project, 46
OnQ, 32, 42
Open Directory Project, 10
OUT, 36, 40, 123
Out of the Past: 400 Years of Gay History, 86
Out4BizUSA, 151
Outing Yourself, 162
Oxenberg, Christina, 20
Oxygen Media's Thrive Online, 141

Pagan spirituality, 97
Parents, 74
Parents, Families, and Friends of Lesbians and Gays (PFLAG)
  description of, 156
  references to, 41
Partner's Task Force for Gay and Lesbian Couples, 89
Pathfinder: Lesbians and Health Care, 136
Peace Corps, 40
Pentimenta, 115
Peplau, Anne, *xiii*
Perry, Troy, 93
Phelps, Fred, 97
Philosophy, 129
PlanetOut
  description of, 24, 162
  interview with CEO, 34-44
  references to, 2, 23, 25, 112
Playing Safe with Both Teams, 136
PLUSnet, 46
Poets and Writers, Inc., 158
Political science, 101, 103-105
The Poly List, 81
The Poly Page, 77
Polyamory
  frequently asked questions, 77
  references to, 10, 55, 71, 72
  resources, 77
Polyamory.com, 77
Popcorn Q, 35
Presbyterian, 93
Press Pass Q, 163
Pride at Work, 149
Pride Events, 155
Project Inform, 140
Prostitutes' Education Network (PENet), 79
Psychological Perspectives on Human Sexuality, 109
Psychology, 101, 105-108
PubMed, 142

QSTUDY-L, 54
Queer Arabs, 99
Queer Arts Resource
  description of, 116
  inteview with founder, 117-125
  references to, 2, 115, 157
Queer Cultural Center, 87, 115, 116, 157
Queer Dimensions, 97
Queer in America, 162
Queer Info Server, 41-42

Queer Jihad, 96
Queer Nation, 45, 46
Queer Resources Directory (QRD)
description of, 27, 28
references to, 2, 25, 40, 112, 118, 162
resources on QRD, 29, 150
workplace listings, 149
Queer Spirits, 90
Queer studies
description of, 49-50
programs in Canada and the United States, 52, 117
resources, 50-54
Queer Usenet newsgroups, 11
QueerNet
description of, 29
interview with founder, 44-48
references to, 2
QueerTheory.com, 53
Quilter, Chris, 144

Radical Faeries, 89, 90, 96, 97
RadioMail Corporation, 30
Rainbow Query, 11, 37
Rauch, Jonathan, 26
Real Networks, 36
Reclaiming History, 86
Reconciling Congregation Program, 93
Religion and Homosexuality, 93
Religious studies, 89
Religious Tolerance, 89
Research Guide for Gay and Lesbian Studies, 53
Rethinking AIDS, 140
RFD, 97
Rhizome, 119
Rielly, Tom, 35
Riggle, Ellen D. B., interview with, 110-113
Right-wing hate groups, 97, 98
Robertson, Pat, 41
Robison, Vanessa, 144
Rosario, Vernon, *xiii*
Roscoe, Will, 90
Rothenberg, David, 165
Roxane Laboratories, 144
Ruben, Gayle, 20

Sacred Sexuality, 10
Safer Sex Pages, 78
Salon.com, 20
San Francisco AIDS Foundation, 140
San Francisco International Lesbian and Gay Film Festival, 34
San Francisco Public Library, 51
San Francisco Sex Information, 78-79
Sanders, Steve, 103
Sandler Communications, 144
Sanlo, Ronni, 14, 51
Sappho, 29
Sarton, May, 86
Sawyer, Tom, 144
Scanlon, Hale, 22
Scanlon, Jessie, 22
Schaub, Kevin, interview with, 156-159
Schüklenk, Udo, 109
Schwartz, L. H., 19
sci.med, 142
sci.med.aids, 142
Search engines and directories
description of, 5-11
use of algorithms, 6
use of Boolean operators, 6
use of metasearches, 7
use of spiders, 6
Sex work
mailing list, 81
references to, 49, 71, 72
resources, 79-80
Sex Workers Alliance of Vancouver, 80
Sexual health and education, 78
Sexual Identity on the Job, 110
Sexual politics, 10
Sexual Science, 74
Sexuality Information and Education Council of the United States (SIECUS), 73, 74
Shilt, Randy, 158
Signorile, Michelangelo, 162
Silverstein, Sharon, 148
Sisters of Perpetual Indulgence, 90

Smith, Gwendolyn Ann
interview with, 66-70
references to, 62
Smith, Megan
and the development of the Internet, 5
interview with, 34-44
references to, 24
soc.bi, 14, 59
The Social Life of Information, 165
Society for Human Sexuality, 74
Society for the Psychological Study of Lesbian, Gay, and Bisexual Issues, 106
Society for the Scientific Study of Sexuality (SSSS), 74
Society of Janus, 77
Society of Lesbian and Gay Anthropologists (SOLGA), 102, 103
The Sociologists' Lesbian, Gay, Bisexual, and Transgendered Caucus of ASA, 108
Sociology, 101, 108, 109
soc.motss, 14
soc.subculture.bondage-.bdsm, 80
soc.support.depression, 142
soc.support.transgendered, 14
Solomon, Alan, 53
Soulforce, 90
Sowadsky, Rick, 134
Sprint PCS, 30
St. Mark's Cathedral (Seattle), 94
Stazer, David, 41
Straight Jobs, Gay Lives, 148
Straight Talk About Gays in the Workplace, 149
Stryker, Susan, 166
Sulayman X, 96
Sullivan, Andrew, 26
Sullivan, Louis, 166
Sunnah, 99

Tadlock, Barry, 110
Talk City, 15
Tantra, 91
Tarver, Chuck, 87
Taylor, Robert, 94
Taylor, Todd, 21
Teena, Brandon, 68
Teens, 74
Telnet, 18
10 Percent, 150
Testicular feminization, 137
The Times of Harvey Milk, *xvii*, 158
Traditional Values Coalition, 98
TransFaith, 94
Transgen, 65
Transgender and intersex health issues, 137
Transgender Education Network, 137
Transgender Forum, 65
Transgender Gazebo, 66, 69
Transgender Law and Policy, 65, 129
Transgender/transsexuals
references to, 49
resources, 61-66
Transsexual Women's Resources, 138
Travis, Terry, 143
Trikone, 88
TS Roadmap, 69
Tufts University, 111
Twice Blessed: The Jewish GLBT Archives Online, 95

uk.rec.motorcycles, 14
Uncle Donald's Castro Street, 158
UnCover, 28
Undergear, 119
United Methodist, 93
United States Department of Defense, 5, 45, 107
United States Department of Health and Human Services, 139, 141
United States Food and Drug Administration, 141
University of California, Berkeley, 32, 41, 165, 166
University of California, Irvine, 31
University of California, Los Angeles, 30
University of California, San Diego, 53
University of California, San Francisco, 139
University of Delaware, 87
University of Illinois at Chicago, 86
University of Kentucky, 110, 111
University of Minnesota, 35

University of Pennsylvania, 117, 123
University of Southern California, 85
University of Texas at Austin, 88
Usenet Newsgroups
  description of, 13, 14
  references to, 18, 30

Varnell, Paul, 26
Veronica, 16
Victory Magazine, 148
VMware, Inc., 44
Volano, 30, 32

Wagner, Wynn and Rick, 144
Walker, Janice, 20, 21
Warhol, Andy, 120
Warren, Patricia Nell, 25
Washington (DC) *Blade*, 31
WebMD, 141
What Sexual Scientists Know, 74
White, Dan, *xvii*
White Crane Journal, 90
White-Olson, Anne, 144
Whosoever: An Online Magazine for Gay, Lesbian, Bisexual, and Transgendered Christians, 94
Wicca, 89, 97
Wide Area Information Servers (WAIS), 16
Wilde, Oscar, 116
Wired. See Citations
Wockner, Rex, 25, 163
Wojnarowicz, David, 123
Women's Funding Network, 154
Workplace discrimination, 149, 150
World Congress of Gay, Lesbian, Bisexual, and Transgender Jewish Organizations, 95
The World History of Male Love, 86
World Sex Guide, 80
WURD Internet Tutorial, 16

XTALK conferencing, 44

Yahoo
  description of, 11
  references to, 15, 36, 41, 50
Yahoo Groups
  description of, 13
  transgender online community, 66
Yale University
  Online Research Guide, 53
  references to, 53, 73
Yang, Jerry, 42

ZapHealth's LGBT Health Issues, 133
The Zuni Man-Woman, 90